Lawless Universe
Science and the Hunt for Reality

JOE ROSEN

The Johns Hopkins University Press

Baltimore

© 2010 The Johns Hopkins University Press
All rights reserved. Published 2010
Printed in the United States of America on acid-free paper
9 8 7 6 5 4 3 2 1

The Johns Hopkins University Press
2715 North Charles Street
Baltimore, Maryland 21218-4363
www.press.jhu.edu

Library of Congress Cataloging-in-Publication Data

Rosen, Joe.
    Lawless universe : science and the hunt for reality / Joe Rosen.
    p. cm.
    Includes bibliographical references and index.
    ISBN-13: 978-0-8018-9580-7 (hardcover : alk. paper)
    ISBN-10: 0-8018-9580-4 (hardcover : alk. paper)
    ISBN-13: 978-0-8018-9581-4 (pbk. : alk. paper)
    ISBN-10: 0-8018-9581-2 (pbk. : alk. paper)
    1. Objectivity. 2. Science–Philosophy. I. Title.
    Q175.32.K45R67 2010
    501–dc22   2009046114

A catalog record for this book is available from the British Library.

*Special discounts are available for bulk purchases of this book. For more information,
please contact Special Sales at 410-516-6936 or specialsales@press.jhu.edu.*

The Johns Hopkins University Press uses environmentally friendly book materials,
including recycled text paper that is composed of at least 30 percent post-consumer
waste, whenever possible. All of our book papers are acid-free, and our jackets and
covers are printed on paper with recycled content.

For Mira

CONTENTS

PREFACE

THE ACHIEVEMENTS OF SCIENCE, and especially of physics, since the start of the twentieth century have radically changed the conception of nature we had held in earlier times. These achievements have revolutionized our view of the universe, even of reality itself. We surely enjoy many of the technological benefits accruing from these advances of science, and also suffer from some of the results. But it is on our intellectual grasp of the world around us and on our role in it, even on our philosophy of reality, where the impact of modern science is truly shattering. As it happens, this intellectual adventure of humankind is largely confined to the circles of scientists and of science-oriented philosophers. And that is a great pity. The general public, who pays for this exciting enterprise, should be given the opportunity to share it.

Fortunately, a number of scientists, who possess the all-too-rare talent of being able to explain their fields to those willing to make the effort to understand, have written popularizing books in recent years. Among them are scientists who themselves play active, even pivotal, roles in the advancement of modern science. Moreover, a number of excellent science writers, who have learned their science well, have joined them. Together, they have produced a wide selection of remarkable books. These books present, in a style accessible to the curious layperson, the modern scientific conception of the fundamental aspects and phenomena of nature, as well as the impressive achievements of scientists in gaining understanding of those aspects and phenomena. Such books are most welcome. All praise is due the authors' attempts to share the excitement with those for whom science is not their way of life.

I hold a serious reservation, however. It seems to me that the lay reader is too likely to gain the impression that not only is science capable of attaining full understanding of the material universe as a whole, in all its aspects and with all its phenomena, including the role of *Homo sapiens* in it, but that science is on the verge of doing so. Some of these authors appear to believe that themselves. They seriously consider the possibility of a "Theory of Everything," capitalized just like that, and occasionally reduced to the acronym TOE. One author even declares that we might "know the mind of God"! Others don't commit themselves, but allow that impression to develop.

Science, by its nature and structure, cannot in principle comprehend the material universe *as a whole*. Through science we can, and indeed do, discover order and laws governing various aspects and phenomena of the material universe, and we gain understanding of them. But the material universe *in its entirety* intrinsically lies beyond science. Within the framework of science, the concept of a law governing the universe as a whole is meaningless. An understanding of everything, a "Theory of Everything," even only of everything of material nature, is but a mirage. As far as science is concerned, the material universe *as a whole* is orderless, lawless, and unexplainable. Any understanding of the whole, if it is to be found at all, can come only from outside science, from nonscientific modes of comprehension and insight.

The main theme of this book is that science, its phenomenal insights notwithstanding, can go only so far in comprehending nature. In this book, we'll see just how far science can go. We'll consider what the domains of science and of metaphysics are, where science leaves off and metaphysics sets forth. Yet we'll observe that science and metaphysics are intimately entangled: metaphysical considerations are unavoidable in science, and science influences our metaphysical positions. We'll learn what science tells us about reality. One thing it tells us is that the objectivity it seeks is not to be found at the level of the world of our personal perceptions and must lie in a deeper reality, one that science does not fully comprehend and likely never will.

The book begins with a discussion and comparison of objectivity and subjectivity. I have come to realize that many human difficulties, even tragedies, stem from an overemphasis of the subjective and a confusion between the subjective and the objective. It is objectivity that we seek through science.

Another theme of the book is an apparent contradiction, even a paradox. On the one hand, the universe *as a whole* is intrinsically orderless, lawless, and unexplainable through science. Yet on the other hand, it is obvious that *within* the universe there are order and laws that science comprehends and explains. How can there be both order and law within the lawless universe?

This leads us to, among other things, the anthropic principle—the use of the existence of human beings as an unconventional explanation of other aspects and phenomena of nature. Following a thorough consideration of its advantages and limitations, we'll see the true status of the anthropic principle as a *scientific* explanatory tool.

Yet another thread is the position of humankind in the universe and in science. That thread leads us to the fact that science is a *human* endeavor. It also leads us to the anthropic principle.

ℐ

This book is a sequel to, a complement for, even an antidote for, books that create the misimpression that science can explain everything. I hope you have already read one or more of them. (Reading them would be useful here, but not essential.) But if you have indeed formed the mistaken impression I'm concerned about, I hope to remedy that. In any case, I hope you'll find that what I have to say is interesting.

My approach is to start from the rather basic definitions and concepts underlying science and to develop them—through simple logic and common sense—to the conclusions I want to convince you of. Since I'm a theoretical physicist, my approach and point of view are, not surprisingly, those of a theoretical physicist. Physics, the most fundamental of the natural sciences, is my model of a science. Other natural sciences differ from physics and from each other, not only in their subject matter, but also in their methods and fundamental conceptions. Yet there is sufficient basic similarity and overlap among all these endeavors to justify my using the term *science* throughout the book to characterize natural science in general.

In this book I present my personal view of science, nature, the universe, reality, and the position of humankind in all that—a sort of The Way Things Are According to JR, or How Things Go According to Joe, or Rosen's Ruminations on Reality. I have my own opinions and biases and don't hesitate to express them, although I do try to put them in perspective and express them clearly. Some of my views are mainstream, shared by most scientists. Others are more or less unconventional, yet fully justified and completely respectable. Among the unconventional is the view that cosmology reaches beyond the domain of science and is basically metaphysics. Another is that the origin of the lawful behavior of quasi-isolated systems lies with the universe as a whole (the so-called extended Mach principle). A third is that the anthropic principle can provide valid, though limited, explanations within science. For more, and for the details, you'll have to read the book.

To the philosophy-minded reader I should emphasize that this is not a philosophical treatise. I'm writing about science and its near neighborhood. I try to present relevant issues as a scientist, not as a philosopher. If I use certain terms and concepts loosely by your standards, I do so to help the presentation and clarify the issues: I'm convinced that for my purposes the gain from so doing outweighs the loss.

Many thanks to my editor, Trevor C. Lipscombe, who is editor-in-chief of the Johns Hopkins University Press. His insightful comments and suggestions have helped me greatly to improve my presentation. And many thanks also to William W. Carver, my copy editor, who made this book much more readable than the manuscript was before he worked his magic on it.

Now let's get down to business and see how science operates within a lawless universe, and how science serves us in our quest for an objective understanding of nature.

Lawless Universe

CHAPTER ONE

Objective or Subjective

That Is the Question

Objective and Subjective

SURELY FEW OF US believe that the world will cease to exist when we die, that everything we perceive is but a figment of our imagination. Such supreme egocentrism, termed *solipsism* by philosophers, holds that the believer has the only "true" existence, and that everything else is merely the believer's mental creation.

As absurd as solipsism might seem at first, it's not that far from the truth. When we see a tree, or think we see a tree, we have become conscious of some mental construct we have learned to call "tree." In the usual course of events, that comes about when light reflected from a tree enters our eyes and stimulates the retinas, whereby electrochemical signals are transmitted via the optic nerves to the brain. The brain then reconfigures itself and we become conscious of "tree."

But does our perception of "tree" mean there is a real tree out there? Not necessarily. It could be only a picture of a tree. Or perhaps a chance confluence of light and shadow has given us the impression of "tree," until we unravel the scene and realize it's nothing of the sort.

Or perhaps there's no external stimulus at all. We could imagine a tree. An artist might do that in great detail. Or we could dream a tree. Sometimes an internally generated "tree" perception might seem more real than the perception produced by a real tree. Or we might call up a memory of a tree. But are we then remembering a real tree or an imagined or dreamed tree? (That sort of thing can be a serious problem, when one's recollection serves as important evidence in a criminal trial.)

So our awareness of the outside world is not at all direct. It is mediated by the same mechanism that produces perceptions that have no external correlates, such as an imagined tree. Then how can we be so sure there *is* a real world out there? Or, put in terms that will be useful for our discussion, What is objective, what is subjective, and what is the difference?

The second edition of the *Random House Unabridged Dictionary* gives a number of definitions for "objective." The one that is relevant to our considerations is

> **objective,** *adj.* existing independent of thought or an observer as part of reality.

I would reformulate this somewhat more clearly as

> **objective,** *adj.* existing as part of reality, independent of thought or of an observer.

The relevant definition for "subjective" given by the same dictionary is

> **subjective,** *adj.* existing in the mind; belonging to the thinking subject rather than to the object of thought.

This definition is good enough as is. The two terms are foundational for our examination of science.

Is the world I perceive, including the computer I'm using at this moment, and even you, the readers of these words, objective? Or is it all subjective? And if subjective, am I the thinking subject or am I but a thought of some other's mind?

As for my being someone's thought, I've been told that René Descartes's (1596–1650) immortal statement, "Cogito ergo sum" (I think, therefore I exist), is supposed to dispel that possibility. Since I experience the sensation of independent thought, my existence apparently can't be merely a role in the production of another mind. I'm not sure I buy this reasoning completely. But since I'm not a philosopher, I'll proceed with my gut feeling that indeed I'm not a figment of another's imagination and will thus accept Descartes's statement.

But what about the computer and you other people out there? Are they all *my* production? And you readers, don't try to tell me you're thinking too. That won't work, since your declarations of independence could also be my imagination's figments. So is it solipsism that we have here, or objectivism? How to resolve this "ism" impasse?

It turns out there's no neat, logically compelling resolution. If I believed I alone am real, and am therefore responsible for the subjective existence of everything else, there would be no logical way to disabuse myself of that belief. (And who are you to argue with your creator anyway?) Still, even though solipsism is a logically unassailable position to take, there is some evidence for a real world: *When we push against the world, it pushes back.*

What I mean is this. If solipsism were valid, I could imagine any kind of world I pleased. But only when I imagine a narrow range of possible worlds do I find myself reasonably pain-free and happy. Otherwise, I would be uncomfortable, to say the least. For example, as long as I imagine that people fall when they jump off roofs, accelerate as they fall, and hit the ground painfully, I avoid jumping off roofs and do reasonably well. But were I to imagine I could jump off the roof with impunity, and were I then to go ahead and jump, I would not be happy with the result.

With sufficient contortion, this situation can be reconciled with solipsism. But is it not simpler and more efficient to allow the existence of a real world? Well, *I* think so, in any case. So I graciously choose to grant you all objective reality and thus abdicate my grave responsibility for your existence and well being. And I assume you choose to reciprocate and relate to me in the same way.

With that finally out of the way and myself off the hook, my computer and I, and you and I, can now face each other as equals in objective existence. We can all—you, my computer, and I—now proceed to discuss the outer, objective, real world and our diverse inner, subjective worlds.

The Objective Outer World: Reality

We now take for granted an external reality, an objective world, a world whose existence and properties are independent of our thoughts about it, and indeed independent of our existence. Before we were born and after we die, the objective world was existing and will exist, with very much the same properties as it possesses now.

What are its properties and how do we know them? Here matters get a bit sticky, because we have no direct knowledge of the objective world. What knowledge we possess, or think we possess, is obtained only indirectly, since we perceive the outer world via our physical senses. And since our perceptions are no more than the stuff of our minds, we can't avoid the conclusion that our knowledge of the objective world is subjective.

Well, isn't that beautiful! What's the point of the objectivity of the real world, if all of our knowledge of it is subjective? We may as well return to solipsism, it would seem. Yet, there is a way out of the impasse. First, we human beings can grant each other objective existence. Second, let's assume we are similar in the way we sense and perceive the real world. If so, we can discuss what we perceive via our senses and reach a consensus about it. As we do so, we constantly perform reality checks by pushing against the world and observing how it pushes back.

The result is a consistent picture, one that is universally accepted among humans, of the real world in which we exist. Technically, this picture is called *intersubjectivity*. It is a consensus obtained from the collective of our subjective knowledges. Our knowledge of the objective world is intersubjective. If, then, we are deluded about the real world, we all live under that *same* delusion. We seem unable to come closer to knowledge of the real world.

But that knowledge we share cannot be far off the mark. If it were, we would not survive too well, or at all. We can invoke Darwinian evolution (Charles Robert Darwin, 1809–1882) to bolster our confidence in our intersubjective knowledge of the world. As humans evolved, those whose perceptions of the world were farther from reality must have tended to fall behind those who perceived the world more nearly as it is—at least in those aspects important for survival. Here we are, the result of all those millennia of evolutionary filtering, thriving reasonably well. So we can't be perceiving the world so badly after all.

Using the evolution argument, one might be tempted to dispense with the need for consensus. If evolution has weeded out gross misperceptions, should not the individual knowledge of each of us, subjective as it is, be good enough without our socializing about it? I'm afraid not. Adaptive evolution involves individual variation. We still must reach consensus.

Intersubjective knowledge of the real world is not objective. Yet intersubjective knowledge *is* independent of the thought and existence of any one individual. So to this extent it has something of the character of objective knowledge. And if intersubjective knowledge is the closest we can come to objective knowledge, we will, for the purpose of this book, simply call our intersubjective knowledge of the world "objective knowledge." Thus we say, with some caution, that through our physical senses, and by means of social interaction, we obtain objective knowledge of the real world.

This conclusion is not free of danger. Intersubjective knowledge might be mass delusion. Recall the Salem witch trials. The resort to such trials was far worse in Europe during medieval times, for everyone *knew* that witches really existed. It took time and unnecessary suffering and death before a reality check put an end to that "knowledge."

One question that might arise is: What about other species? What about birds, dolphins, chimpanzees? Do they perceive the world as we do? There's no way for us to know until we develop far more sophisticated communications with these animals than we have at our disposal now. By the evolution argument, their perceptions, too, should not be badly out of tune with reality. And since the aspects of reality that are important to them overlap to some extent those that are important to us, we can reasonably assume their perceptions have certain similarities to ours.

There are of course aspects of reality that are important to animals and not to humans, and vice versa. So their perceptions will certainly differ from ours in major ways. For example, dolphins can precisely sense water pressures and currents along their bodies and make great use of these perceptions. And in addition to their vision, dolphins use underwater sonic echolocation to "see" very effectively. These aspects of reality are hardly accessible to our senses.

What might we expect from extraterrestrial creatures? The evolution argument would be just as valid for them. So we would expect to see much the pattern we see in the animals: domains of overlap and of difference. For instance, if their home planet happened to revolve around a hotter star than our sun, they might see light that is in the ultraviolet range and invisible to us. But then what we see as red light might be invisible to them.

Now if the social aspect is so important for our objective (actually intersubjective, don't forget) knowledge of reality, then might our knowledge of reality be a social convention or construction? There are those who claim this is the case, along the lines of postmodernism and deconstruction in the humanities. If true, then female knowledge of reality might differ from male knowledge, Western knowledge might differ from Eastern, and so on.

The deconstructionists claim, if I have it correctly, that art, literature, and so forth have no intrinsic meaning, and that only their viewers, listeners, and readers give them meaning, in a social context. That makes some sense to me, but I wouldn't deprive the author or artist of any voice in the matter, as I believe the deconstructionists, or perhaps only the most extreme among them, do. Still, it became increasingly fashionable for a time, and therefore common to claim that our objective knowledge of reality is also a social convention.

Put bluntly and undiplomatically, that is nonsense. Knowing reality is not the same as understanding a piece of literature or appreciating an artwork. First, there is the hardness of reality. It's not malleable. When we push it, it pushes back. You fall and get hurt when you jump from a roof—whether you're female or male, Eastern or Western—and no amount of contrary social deconstruction will save your bones from breaking. That is the reality check.

If you choose to interpret a story, sculpture, or symphony one way or another, you're perfectly free to do so. The work itself doesn't have the power to object to your interpretation or correct it (although the author, artist, or composer might). And if different people interpret the same work differently, so what? Even if your interpretation flies in the face of the author's, artist's, or composer's intent, what does it matter? The work itself accepts any interpretation by anyone. Anything goes.

Knowledge of the *real* world is a different matter altogether. If your "knowledge" passes the reality check, it is indeed knowledge. If my "knowledge" fails the reality check, it's simply not knowledge. Although there is room for interpretation in our knowledge of reality, the reality check imposes objective constraints on interpretations. It's not true that anything goes. Much does not.

As a fanciful example, interpret the manner of falling off a roof. Perhaps, correctly, there's a strong attractive influence from Earth. Or instead, there might be a repulsion from the atmosphere. Or maybe the Moon attracts, or the Sun repels. But here, as opposed to literary, artistic, or musical interpretation, not every point of view is as valid as any other, and most views aren't valid at all. Here there is the reality check. If you jump from the roof, you follow an approximately parabolic trajectory that ends in a thud on the ground. If your interpretation isn't consistent with those facts, it's invalid.

Then there's the evolution argument. If you don't perceive reality sufficiently closely to what it really is, you won't have great evolutionary fitness, and you won't survive to pass your genes on. The reality check again; there is a "really is." We are not a society of roof jumpers. Jumping from roofs is objectively dangerous. Those who are inclined toward roof jumping, considering it to be safe and pleasant, remove themselves from the human gene pool and don't pass their inclination on to later generations. So we remain a society containing very few roof jumpers.

Yet, knowledge of reality as a social construct is not altogether utter nonsense. The domains of reality that are important to different human beings and to different human groups overlap to a great extent, since we're all basically similar. But those domains don't necessarily overlap totally. So different groups, and even different individuals, might have somewhat different knowledge of the same aspects of reality. As an example, my perception of wind strengths is probably described by the scale "still air, breeze, medium wind, strong wind, devastating wind," since these rough distinctions are all that affect me. But operators of sailboats surely make much finer distinctions, ones that are important for their safety and survival.

I stated above that our objective (really intersubjective) knowledge of reality is obtained through our physical senses and by means of social interaction. But who participates in the social interaction? Who takes part in the consensus? Clearly, we won't ask the blind to help with issues of vision, nor will we expect the deaf to contribute to auditory matters. And we won't ask the raving lunatic to take part at all. But where do we draw the line? I'm afraid I can't pin it down. I suppose we're talking about a consensus of a sufficiently large majority of people, whatever "sufficiently large" might mean. There will always be some minority who don't participate, or who hold firm to a dissenting opinion. We'll have to live with this ambiguity.

As an example, imagine a flat object standing upright on the village green. We ask passersby to view the object from all angles and describe its shape. Say 98 percent report the object is square. Of the other 2 percent, let half describe it as rectangular with its height greater than its width (tall), and let the other half perceive it as a rectangle with height less than width (squat). As a result, our intersubjective determination is that the object is really, objectively square, and the dissenting 2 percent are presumed to have problems with visual perception. Then, as a reality check, we measure the object, with great care, and confirm our determination.

What if our measurement proved the rectangle was indeed objectively squat? That would force us to qualify our determination and come up with an explanation for the optical illusion, to which 1 percent of viewers were immune.

Why should our objective knowledge of reality be obtained only through our physical senses? I claimed above that we have no direct knowledge of the objective world. Is that in fact the case? Some people do claim direct knowledge of reality. Recall those who claim to be able to locate lost or kidnapped children and other missing persons. And what about ESP, clairvoyance, and the like? All that makes great material for the sensation tabloids we laugh at, or should laugh at, when we pass them at newsstands or at the supermarket checkout.

To the best of my knowledge, carefully controlled tests have invariably disproved any claim of direct knowledge of reality. No, it is only through our physical senses that we perceive the objective world. For humans, these include hearing, sight, taste, smell, temperature, gravity, force, touch, and still others (forget the classical "five senses"). For every such sense there are identifiable physiological sensors in the body. And every such sense has to do with an identifiable physical effect. For sight, the sensors are the eyes, whose most critical components are the retinal rods and cones, which respond to that range of electromagnetic radiation we call light.

In summary, then, we accept the existence of an objective real world, which exists independently of our observing or thinking. We gain knowledge of it through our physical senses and by means of social interaction. This knowledge is intersubjective, but for the purposes of our discussion we call it objective.

And going beyond knowledge, what about understanding? How should we attempt to make sense of the objective knowledge we accumulate about the real world? Objective knowledge calls for objective means for attaining understanding. Those means are science. More, much more, on science—starting with the next chapter.

Our Subjective Inner Worlds: Fantasies

Let's leave, for now, the outer, objective world—the real world—and our knowledge of it and turn to our subjective worlds, the inner worlds of our minds.

The stuff of our minds, as we have seen, includes what we call objective knowledge of the real world, although this knowledge is intersubjective and derives from subjective perceptions resulting from our stimulated physical senses. We categorize the rest of our "mind stuff" as our subjective worlds, our inner worlds, our fantasies. Whereas our knowledge of the outer world is indirect, we each have immediate and direct knowledge of our own individual private inner world. And this knowledge is so private that it is completely inaccessible to any other person.

If that isn't clear, think about this. Imagine your perception of the color of the clear daytime sky. We agree to call this color "light blue" or "sky blue"—some shade of blue. Can you describe to another person the perception of blueness you have? No way! After mumbling a bit, you might come up with: "Well, just look at the sky. *That* is what I perceive." But you've conveyed nothing. You haven't, and can't have, the slightest idea whether what your friend calls "blueness" and perceives while looking at the sky is anything like your own perception.

That is the essential nature of our inner worlds. The subjective fundamentally cannot be revealed to others. So our various individual subjective worlds are incommensurable; they have no common measure. As with our perceptions of blueness, there is no way to compare yours with mine.

"Yet," you might object, "surely I can say to someone that I see blue, and that person will understand what I'm talking about." Yes, and you can say you're hungry and that person will also understand what you're talking about. That's due to language conventions. Each of us learns to associate words with certain perceptions, feelings, etc. As long as the perceptions and feelings are somehow correlated with the real world, we take it for granted that others have similar perceptions and feelings when they're called by the same name.

"Hungry," for example, has to do with not having eaten for some time and badly wanting food. So when I tell you I'm hungry, you have some idea what I'm talking about. But we still can't compare our respective feelings of hunger. They're purely subjective.

For another example, "sad" has to do with being in situations that are generally agreed to be undesirable, such as losing in the stock market or hearing about the accident of a friend. So when you tell me you're sad, I do have some idea what you're talking about. And yet there is absolutely no way we can compare our respective feelings of sadness. Again, they're purely subjective. And so on and so on—through any number of examples of subjectivity.

The incommensurability of our individual inner worlds is an inescapable fact of the human condition, and each of us is deeply alone in this respect. What should we then do? Withdraw from the world and hide inside our shells? I should hope not! We can and do still try to reach out and communicate. We talk endlessly. We write literature and poetry. We compose and perform music and create visual artwork. We use analogy, simile, allegory, metaphor, parable, and association. We make the most of shared experiences and common language. We even learn second or third languages.

Perceptions, feelings, emotions, spirituality, and the like are subjective. To this list I now add belief, which is an internally held conviction. It exists in the mind and belongs to the thinking subject rather than to the object of the belief. It is therefore subjective.

I don't expect to have to do any convincing about our more or less arbitrary beliefs: This time I'll win the sweepstakes. The tossed coin will come up heads. I'm about to be dealt three aces. It will rain tonight. I'll beat your tail off in Monopoly. Just because you believe it, it doesn't have to be so.

But what if belief involves potentially predictable aspects of the real world? Is it then still subjective? Yes. It has to be, as we just saw. Furthermore, consider this example. On the basis of the stock market's past behavior and our individual understanding of its behavior, you predict it will rise tomorrow and I predict it will fall. (That's why there are winners and losers in the stock market.) Since beliefs about even the real world can differ, as in this example, again there can be nothing objective about belief.

But what if beliefs about the real world do not differ? Consider this case. Say you believe the Sun will rise tomorrow. And say you achieved that belief by careful consideration of your past experience, the recorded experiences of others, and your understanding of astronomy. Yet, someone else could in principle reach the conclusion—from the same or similar data—that your belief is unwarranted, that the Sun might not rise tomorrow.

Such disagreement, although possible in principle, does not occur in practice for the case of the Sun's rising. On the basis of past experience and our understanding of the Sun's apparent motion in the sky, we all agree, without hesitation, that the Sun will rise tomorrow. We would bet anything on it. So we all hold the same belief about the Sun's rising tomorrow.

Does that not make the belief objective? Still, no. It does, however, make the belief intersubjective, since we're all in agreement about it, so it is independent of the thought and existence of any individual. And in this sense, and only in this sense, we *agree* to call such a belief objective. In fact, we're back to our "objective," but really intersubjective, knowledge of the real world. The prediction of the Sun's rising tomorrow is knowledge of the behavior of the real world.

As an aside, how could anybody believe the Sun might not rise tomorrow? The answer is that it might not! That has to do with the way science works, which we discuss farther on. But in brief, it's a matter of likelihood. Our knowledge of the real world with regard to the Sun's rising is great. But no scientific knowledge is absolute; it's always subject to new discoveries and improved understanding.

So what we can say in full disclosure is that, in light of our present understanding, it appears very, very likely the Sun will rise tomorrow, so likely that we're willing to bet the farm on it. But the Sun just might explode overnight and blast the Earth from its orbit.

Our present understanding of stellar evolution precludes that happening, at least for many millions of years. But how good is our present understanding? (Hope you have a good sleep tonight.)

We can compare this objective, but really intersubjective, knowledge about Earth and the Sun to similar knowledge in earlier centuries. Then everybody *knew* the Sun rose and set as it revolved around central Earth, and *knew* the other planets and the stars also revolved around Earth. All the evidence, such as it was, seemed to point to that. And—of much greater importance for most—religious dogma supported the picture. The geocentric universe was the objective (really intersubjective) knowledge of those eras. That "knowledge" has long since undergone reality checks and failed them. Here we are with our present-day objective (really intersubjective) knowledge of the heliocentric solar system and the place of the solar system in our galaxy, our galaxy in its galaxy cluster, and so on. This knowledge has passed reality checks. So far. But what will people think of our "knowledge" a hundred years from now?

Objective or Subjective?

The impression might be growing that I'm about to claim everything is either objective or subjective. Not quite. We will eventually reach a point in our discussion where I'll be pushing for more carefully distinguishing between the objective and the subjective. But as for their being mutually exclusive categories, the situation is far from clear.

In the "subjective" category we have so far accumulated perceptions, feelings, emotions, and the like, to which we added belief. We saw that belief about the real world, if it's universally agreed to and thus intersubjective, has something of the objective about it, since it is independent of the thought and existence of any individual.

Now, what about abstract logical structures, including mathematics? Take as a simple example the mathematical truth expressed by the relation $2 + 3 = 5$. Mathematics is a self-consistent abstract logical structure involving definitions, axioms, operations, and so on.

Some of mathematics is applicable to the real world. (One might well wonder why that's so, why a purely abstract structure, the fruit of purely mental activity, should turn out to so aptly describe the real world.) Such is the relation $2 + 3 = 5$, since two apples added to three apples do make five apples. Certain branches of mathematics were developed expressly in order to better describe and understand the real world. Examples are arithmetic, geometry, and calculus.

Other fields of mathematics, however, were first developed in a purely abstract manner, and only later was it discovered how applicable they are to the real world. Such here are complex analysis, group theory, and Riemannian geometry, as examples. And the applicability of yet other branches of mathematics is not presently known.

What is the nature of mathematical truths and, more generally, of logical constructs? Certainly they are at least subjective, since they are mental constructs. Since they're universally agreed upon, they're even intersubjective. But could this intersubjectivity not be a reflection of actual objectivity? Not objectivity in the sense of such constructs being relevant to the real world, but rather in the sense of themselves having objective existence.

The question is: Do logical constructs in any sense have existence independently of our knowing them? Are they actually "out there" and merely get *discovered* by humans, and presumably also by any other sufficient intelligence? In this case we would consider them to be objective. They would then have had some mode of existence even before intelligence ever developed in the universe, and they would still be around in some sense, somewhere, even if all intelligence were to be extinguished.

Or does the existence of logical structures hinge so crucially on mental activity that they are *created*, rather than discovered, by intelligence? Then without thought they would have no existence at all; we would consider them to be purely intersubjective. I have no answer or enlightenment to offer. There are those who hold one view, and there are those who hold the other. And there are those who simply shake their heads in puzzlement. For what it's worth,

I think I would place myself in a centrist-to-objectivist position. For now, at any rate. You smile? Well, what do *you* think?

Objective Truth

In our discussion, we have made use of three kinds of truth, and they are the only kinds of truth I can think of. One kind of truth is consistency with reality. We might then call it objective truth. So a statement, claim, or idea is objectively true if it passes the reality check, if it is consistent with the real world.

For example, the statement "People normally fall to Earth when they jump off a roof" is objectively true. And so is the claim that the Moon revolves around Earth. On the other hand, the idea that I have not yet been born is objectively false, as is the claim that Bill Clinton was the third president of the United States.

Objective truth is a straightforward and obvious kind of truth. We make use of it all the time. You drank coffee this morning, or perhaps you did not. I either did or did not rake my yard yesterday afternoon. There is or is not a giant black hole at the center of the Milky Way galaxy.

Subjective Truth

Another kind of truth has to do with subjective matters, and it can accordingly be called subjective truth. "You felt sad last night" would be subjectively true, if in fact you felt sad last night. But there is no reality check here, no comparison with the real world, since your feelings are not part of the real world. They are part of your private, subjective inner world.

The statement "You *looked* sad last night" could be objectively true, if your demeanor and facial expression last night objectively matched a certain agreed-upon pattern. And statements about your heart rate, blood pressure, hormone levels, breathing rate and depth, neuronal activity in your brain, and so on last night could also be objectively true or false. But your feelings are strictly your own business, and the only way to discover the truth about them is for us to ask you, and you to be truthful in your response.

Similarly, the truth or falsehood of statements about our thoughts is subjective. Our thoughts are not part of the real world. There is no reality check for thoughts. To find out about my thoughts, you must ask me.

Beliefs are subjective truths. If they're universally held, beliefs will also be intersubjective truths. If, in addition, they're reality checked, they'll be objective truths.

Your personal beliefs are your own subjective truths. As strongly as you might hold your beliefs, as important a part of your life as they may be, and as real as your beliefs might be *for you*, as long as they don't pass the reality check against the objective world, they remain solely your own subjective truths. As such, they need have no validity for anybody else. I might hold beliefs very different from yours. I might even think your cherished beliefs are nonsense. But if my beliefs don't pass the reality check either, they remain solely my own subjective truths. Your beliefs might be nonsense *for me*, but I would have no basis for claiming that they're nonsense *for you*. They might be very dear to you. There is no contradiction here, and no argument between us.

Really, no contradiction, no argument? Not over beliefs that are solely subjective truths. What could there be to argue about? Your subjective truths are true only for you, in *your* inner world, while mine are true only for me, in *my* inner world. There's no commonality for our subjective truths, no single world to which both sets of truths apply, so no possibility of clash or contradiction and no argument.

Let's consider an example. Say you believe in the inherent goodness of humanity, while I believe in humanity's inherent evil. Our beliefs sound as if they might reflect the real world, so we run reality checks on them. The results? Negative for both. You do come up with a lot of evidence for inherent goodness, which supports your belief and contradicts mine. But then I present much evidence for inherent evil, in support of my belief and contradicting yours. Since a contradiction negates a reality check, regardless of any positive evidence, neither belief passes.

So is humanity inherently good or inherently evil? The evidence shows that neither is a valid objective description. And wiser heads than ours point out that in any case we first have to carefully define what we mean by "good" and by "evil," determine how they are to be detected, and understand the significance of "inherent" for humanity, among other minor details.

And if, in spite of having our beliefs shown not to be objective truths, we still insist on holding them, do our beliefs then stand in contradiction? It might seem strange in this example, but they don't. Neither belief expresses an objective truth, so both are wholly subjective truths. As such, each belief is true only in the mind of its believer. Put crudely, what that amounts to is this. Humanity is shown to be neither inherently good nor inherently evil. Yet we both persist in fantasizing, each of us in his or her own mind—you, that humanity is inherently good, while I, that it's evil. Each fantasy exists in its own private world, and there's no contradiction at all.

Let's try another example. Imagine that you believe there is one god, whereas I believe there are three. By "god" we agree to mean some powerful supernatural, or transcendent, entity that can interact with the objective real world, can influence it and be influenced by it. By "supernatural" or "transcendent" we mean not belonging to the natural world. And since this isn't a philosophical treatise, let's not involve ourselves in a discussion of what kind of existence that might be.

Once, while flying to attend a conference on science and knowledge, I was studying some preparatory material for the conference and found it was being taken for granted that two such beliefs would be contradictory. That was astonishing, since there is no contradiction at all. Use the reality check. You collect evidence from the real world for one god and against three, and I search for evidence for three and against one. But as we well know, there is none of either to be found.

I hope we do know that well. There's nothing in the real world that supports or contradicts *any* claim about the transcendent or supernatural. If there were, we would then have objective evidence about the transcendent/supernatural, everybody would agree, and there would be no controversy over the matter. That is obviously so far from the case, it's laughable to even imagine it. *The Twilight Zone, The X Files,* and the tabloids notwithstanding, to the best of my knowledge there are no credible alleged cases of supernatural intrusion into the affairs of the real world that haven't been shown to be explainable in real-world terms.

And in particular, for the beliefs in one and three gods, the best thing a referee could do would be to ask us to produce our gods and hold a confrontation. Well, I certainly can't produce my three. If you can produce your one, you win. So go ahead. I'm waiting, and I expect I'll have to wait forever. Our beliefs are simply subjective truths, true for each of us alone. Your single god and my three gods are products of our minds. And their claimed transcendent, or supernatural, existence, as real and important as it might be for us, is also a product of our minds.

Yet, some might insist, why can the transcendent not belong to the objective outer world? Must the real world consist only of the natural world? Is there nothing objective about the transcendent? Let's check. Do we perceive it via our senses? Clearly, no. Perceiving the transcendent through our senses would be an oxymoron. Our sense organs are part of the natural world and are stimulated by identifiable natural effects. So whatever we perceive via our senses can only be natural.

Perhaps, then, we can find the transcendent in the natural world indirectly and holistically, such as in its beauty, its order, its wonder. Are the likes of these inherent to the natural world or are they purely mental constructs? Beauty, wonder, and such—powerful though they may be—are mental constructs. By contrast, order, lawful behavior, and the like can be argued either way. If we're dealing with mental constructs, then we're back to subjective truth. But if we have widespread consensus about something that's invisible but inherent to the natural world, such as order perhaps, then we have intersubjectivity and can call it objective truth.

But is a notion like order transcendent? Is nature's order, for example, taken as inherent to nature, transcendent? We're now down to semantics. Have it your own way.

To continue with the example, if we take nature's order to be inherent to nature, and if we choose to view it as transcendent, do we not then have objective evidence, real-world evidence, about something transcendent? It seems we do. So my earlier statement, that there is nothing in the real world that supports any claim about the transcendent or the supernatural, would appear to be too dogmatic. With "transcendent" sufficiently loosely defined, one can find real-world evidence about the transcendent. To be sure, observations of the real world shed light on nature's order. But after all, with sufficiently loose definitions one can conclude pretty much whatever one chooses.

As for my own taste in the matter, I would prefer to have my dogmatic statement be tautologically true. I would have it that if there is real-world evidence for something, then that something is *by definition* not transcendent. Different people make different choices here. Consensus is lacking.

And gods, one or all, what about them? Do we not have something theistic in the natural phenomena of life, in the complexity and fitness of living things, in the concept of order (taken to be inherent to nature), and so forth? Can theistic beliefs express objective truths? Even then, no. Even if we so loosely define transcendence as to include such natural phenomena, and choose to interpret this transcendence theistically (thus adopting what is known as a natural theology), where is the intersubjectivity that would allow us to declare it an objective truth? Nowhere. Such matters are so far from being intersubjective, and so obviously so, that belaboring the issue is a waste of energy.

The bottom line here is that transcendence, supernaturality, gods, and so on, whether defined tightly or loosely, are not part of the objective real world. That follows from consistent failure of the reality check, from lack of consensus, or both. One's belief in such matters is one's own private subjective truth.

There might be additional believers with similar beliefs. (Since beliefs are subjective, there's no way to reliably compare beliefs of different people. We can only estimate, based on behavior and self descriptions.) There might be vast groups of people holding similar beliefs. And it's at least possible that a large majority of people, worldwide, will eventually come to hold similar beliefs. If that majority is sufficiently great, the common beliefs might even become intersubjective. But does that make the beliefs any more objective than if they're held by only a handful of people? Or a single individual?

By no means. For a belief to be objective truth, there remains the hurdle of passing the reality check. No matter how dear and popular the belief, if it doesn't pass the reality check, there's no way it can lay claim to any shred of objectivity. Consider again the beliefs in witches and geocentricity, as examples.

Logical Truth

The third kind of truth, in addition to objective and subjective truths, is logical truth. This is the truth of logical deductions, including mathematical theorems, within the self-consistent frameworks in which they are derived. (A self-consistent framework is one in which there are no internal contradictions.) For example, this syllogism is logically true: All New Yorkers are rude; Joe is a New Yorker; therefore Joe is rude. The syllogism can be expressed also as: If all New Yorkers were rude, and if there were a New Yorker named Joe, then it is logically true that Joe would be rude.

Logical truths need have nothing to do with the real world; a logical truth does not have to be either objectively true or objectively false. The New York example is valid whether or not all New Yorkers are rude, and whether or not any person named Joe lives in New York. But a logical truth *may* also be objectively true. For example, the logically true mathematical relation $2 + 3 = 5$ possesses also objective validity, since two bananas and three bananas together make five bananas.

Or a logical truth may also be objectively false. Here's an example. A theory, as the term is used in science, is an explanation, or an attempted explanation, of observed phenomena, or of a law of nature. (More about that in chapter 3. Please don't confuse this usage with the common, everyday use of the term *theory* to more or less mean speculation or conjecture.) In science, any theory, since it serves as an explanation, must be logically self-consistent (i.e., must be contradiction free). So any conclusion of a theory is logically true within the framework of that theory.

A good theory is one whose conclusions are all objectively true, and whose claim successfully explains whatever real-world phenomena or law it is intended to explain. But for a theory deemed unsuccessful, at least one conclusion, although *logically* true, will be *objectively* false.

An example of an unsuccessful theory consists of Isaac Newton's (1642–1727) laws of motion (of which more in our discussion of science farther on) and James Clerk Maxwell's (1831–1879) laws of electromagnetism. This is a theory explaining mechanical-electromagnetic phenomena. The theory does an excellent job for phenomena at scales larger than the submicroscopic, and so is a successful theory *at those scales*. But it fails at the molecular scale and smaller scales. For example, it predicts that all atoms are fundamentally unstable and can't exist for longer than a small fraction of a second. Yet here we are, along with all the stable atoms constituting our bodies, thinking about truth. (Quantum mechanics and quantum electrodynamics form a successful theory that encompasses the molecular and atomic scales as well.)

Dealing with the Subjective

Our subjective inner worlds are highly individual. Yet since people are basically similar, our subjective worlds appear to have much in common. We usually don't feel too frustrated when we attempt to communicate something of the content of our inner worlds to others. One might claim that the struggle to overcome the barriers that separate our private inner worlds adds spice to life.

Many people even make a living dealing with the subjective. Psychologists and psychiatrists immediately come to mind. Then there are authors, poets, composers, and artists, to which we can add clergy, religious leaders, prophets, spiritual gurus, televangelists, and advertisers. But let's not leave out palm readers, fortune tellers, astrologers, mediums, and promoters of bible codes.

Popular culture magnifies the subjective to paramount importance. Don't just be in touch with your feelings; let them rule your life. Self-esteem above accomplishment. Suppress objective truth, if it might be unpleasant for some group of people. And religion, alongside popular culture, raises the subjective to the heights of concern.

To a large extent, I feel it is due to religion and popular culture that so many people have an alarming difficulty in differentiating between the objective world and their individual subjective ones, between reality and fantasy.

For babies and infants it's normal that the two worlds be merged seamlessly. Their feelings and desires are their only reality. It's part of the human maturation process that a separation between the inner and the outer be developed. That's a gradual process. Children can be terrified of their fantasized monsters lurking in the dark. And even into maturity, people might tend to assume that others think just as they do. For a truly mature and self-aware person, however, the two worlds are disjunct.

Yet, many people think their subjective truths are universal and objective truths, that only they are the true believers and all others are in error. This attitude is powerfully reinforced by the believer's being a member of a like-minded group.

A case in point is creationism. Let's use the term in a generalized way to denote the projection of one's subjective truth to the distant past: one takes advantage of the fact that in the distant past no one was around to report on the situation. And in the more recent past there were indeed observers, but they left no records, or left records whose objective validity is not universally accepted. So one feels secure in claiming that one's belief about the past (say, divine creation or "intelligent design") is the way things objectively happened. Contrary objective evidence (fossils, carbon-14 dating, astronomical observations, unintelligent design, and so on) is ignored or reinterpreted. Voilà, creationism.

To address creationism, we need to agree that we're talking about an objective real past. Something or other objectively did or did not happen; anybody who was present at that time would report what it was that did or did not happen.

One approach is to claim that the nature of the objective past should be investigated through all objective means available to us. Beliefs and holy scriptures are not objective means. Science is the most nearly objective method we have for dealing with the objective real world. Science has developed a scenario for the past (biological, geological, and cosmological evolutions, for example). The scenario, like the rest of science, is not absolute and not immutable. Uncertainties are certain to arise, but any further uncertainties in the scenario are matters to be resolved through science.

Thus a rational person should best be guided by the scenario proposed by science, even with its contingent nature. And if one nevertheless insists on preferring one's subjective truth to that of science, then, I would claim, to be consistent one should avoid riding in cars, flying in airplanes, and undergoing radiation therapy. Their safety, efficacy, and reliability are based on the same objective reasoning that produces science's scenario of the past.

Here is another approach. Creationists claim to have gained knowledge of the objective past through their beliefs and holy scriptures. But among humans there exist many different beliefs and holy scriptures. So there are many different creationist "knowledges" about the objective past. For example, the Hebrew bible is often interpreted as having the world created some 6,000 years ago. But a version of Hindu cosmology has the world existing forever and evolving cyclically. And even within a single scripture or belief there might be found contradictions. In the first book of the Hebrew bible, for instance, the first chapter relates that the first humans were created after the animals, but the second chapter has the order reversed.

These various descriptions of the past cannot all be the *objective* truth, for they would contradict each other on many counts. (They can all coexist peacefully as *subjective* truths, as indeed they do.) Every creationist claims his or her narrative is objectively true. As impartial observers, why should the rest of us accept one story and reject the others? What is there about one version of the past (such as 6,000 years or forever, people before or after animals) that makes it the objective truth and the other versions fables? The answer is: nothing. None of the versions is based on objective evidence. Their sole support is the (subjective) belief of their believers.

Dealing with the Objective

If the objective real world is to be dealt with in a meaningful way, it must be dealt with objectively. The most nearly objective method we have for comprehending the real world is science. So it is to science we turn, and we discuss its nature and limitations in the following chapters.

We live in an objective real world, and we gain objective (actually intersubjective) knowledge of it through our physical senses and by means of social interaction. Yet our knowledge of reality is not merely a social construct. The reliability of our knowledge derives from reality checks and from our evolutionary fitness.

Each of us possesses his or her own subjective inner world. That is where perception, feeling, emotion, belief, etc. lie. Our individual inner worlds are incommensurate with those of others.

The status of abstract logical structures (including mathematics) vis-à-vis the objectivity/subjectivity dichotomy is unclear.

Objective truth is consistency with the objective real world. Subjective truth lives in the inner world of each of us. Beliefs are subjective truths. Logical truth is the truth of logical deductions (including mathematical theorems) within the self-consistent frameworks in which they have been derived.

Subjective truths are all too often taken to be objective truths. Creationism is a current example of subjective truth being taken to be objective truth.

The objective real world is best comprehended by means of science.

Bibliography

Here are some books that I have found relevant to this chapter, though their authors have different views and take different approaches. For example, Polkinghorne is a physicist turned ordained minister, whereas the physicist Weinberg is an avowed atheist, as was the astronomer Sagan. The latter was the author of the novel *Contact*, which was the basis of the movie with the same name. Shermer, a debunker of pseudoscience, writes a regular column in *Scientific American*. Gould, a paleontologist and biologist, held a broad view of science and wrote extensively. The book edited by Gross, Levitt, and Lewis is a collection of conference contributions, among which a reader is advised to pick and choose. I enjoyed reading all those I tackled and am sure you will too.

A. H. Cromer, *Connected Knowledge: Science, Philosophy, and Education* (Oxford University Press, Oxford, 1997).

———, *Uncommon Sense: The Heretical Nature of Science* (Oxford University Press, Oxford, 1993).

S. J. Gould, *Science and Religion in the Fullness of Life* (Random House, New York, 1998).

P. R. Gross, N. Levitt, and M. W. Lewis, eds., *The Flight from Science and Reason* (New York Academy of Sciences, New York, 1996).

P. Kurtz, ed., *Science and Religion: Are They Compatible?* (Prometheus, Amherst, N.Y., 2003).

R. G. Newton, *The Truth of Science: Physical Theories and Reality* (Harvard University Press, Cambridge, Mass., 1997).

J. C. Polkinghorne, *Belief in God in an Age of Science* (Yale University Press, New Haven, Conn., 1998).

C. Sagan, *The Demon-Haunted World: Science as a Candle in the Dark* (Ballantine Books, New York, 1996).

M. Shermer, *Why People Believe Weird Things: Pseudoscience, Superstition, and Other Confusions of Our Time* (Freeman, San Francisco, 1997).

V. J. Stenger, *Physics and Psychics: The Search for a World beyond the Senses* (Prometheus, Amherst, N.Y., 1992).

S. Weinberg, *Facing Up: Science and Its Cultural Adversaries* (Harvard University Press, Cambridge, Mass., 2001).

CHAPTER TWO

The Science of Nature and the Nature of Science

Preliminaries

WHAT, THEN, IS SCIENCE, which I claim to be the most nearly objective means we have for comprehending the real world? The rest of this book is devoted to answering that question. In the following we'll see what science is and how it operates. We'll discover the objective nature of science. Yet we'll learn why, nevertheless, some degree of subjectivity can't be avoided in doing science. We'll also look into what lies beyond science, into some of the general intellectual framework of which science forms a part. And we'll consider the limits of science, what science can and does deal with and what science cannot deal with.

Many people, among them a large number of scientists, believe science is the fountainhead of understanding of, if not everything, then at least everything of material nature. They are mistaken.

Granted, science is usually not expected to probe or comprehend the nonmaterial, subjective aspects of the universe, such as morals, beauty, love, and faith. Most of us realize that, and we limit our expectations accordingly. (Even so, it appears that considerations of evolutionary advantage can shed light on matters such as altruistic behavior, perception of human beauty, and morality.) And there are aspects of the universe, such as life, consciousness, mind, and intelligence, whose materiality (or lack thereof) is the subject of heated debate. Although many people, and I among them, take a materialist position with regard to these aspects, not everyone thinks science is able to deal productively with such issues.

But even in its handling of the material universe, science, by its very nature, has intrinsic limitations. One such limitation is that it cannot comprehend the material universe *as a whole*. Science does indeed give us understanding of various aspects of the material universe and the phenomena within it. But science can go only so far. In principle, the material universe *as a whole* remains beyond science's grasp.

Nonscientific modes of comprehending and understanding, such as feeling, intuition, and religion, have, with the coming of the age of enlightenment and the rise of rationalism, become largely relegated to those aspects of the universe that lie outside the domain of validity of science. And I fully concur with that. In dealing with all those aspects of the universe that science does successfully comprehend, or is potentially capable of comprehending, nonscientific modes of comprehending should indeed be rejected as sources of objective understanding.

As an example, consider the multicolor effect obtained when sunlight passes through a glass prism or through drops of water. Science offers an explanation of the phenomenon in terms of the wave nature of light, the dependence of the speed of light in glass and water on the frequency of the transmitted light, and so on. Moreover, the explanation offered by science holds through myriad observations and experiments. A nonscientific explanation might be that the spectrum is the direct result of God's will to enrich our lives with the beauty of the colors thus manifested, or is a sign of God's promise to humankind that there will be no repetition of the devastating flood of eons past. The scientific explanation is objective, and we can all agree on it. It is also useful in that it enables us to control the prismatic effect with advantageous results. In this case, the nonscientific explanations do not clash with the scientific one. Because they are subjective, one can take them or leave them. They do not enable any control. I'm ever anew full of wonder and delight at the sight of a spectrum, especially as manifested in a rainbow, and the sight of a rainbow can even remind me of a certain biblical narrative. But it is the scientific explanation that comes to my mind when I am in explaining mode, both for myself and for my students.

Those who prefer explanations involving feeling or belief, rather than scientific explanations, in those domains where science is valid are welcome to them. That's their personal business. But the encroachment of nonscientific modes of comprehension on the domain of validity of science, as a matter of public policy, is another matter altogether. Science, where valid, offers by far the most nearly objective comprehension and understanding available, and is thus the only mode of comprehension suitable for general public recognition. It's a unifying factor for humankind and a firm foundation for a world culture. Nonscientific modes of comprehension, for their part, are subjective. They're best left to the individual, or to like-feeling and like-believing groups of individuals, and should not be allowed to become public policy in any domain where the validity of science has been demonstrated.

Imagine what might happen if a country did base public policy on nonscientific modes of understanding in a domain where science has a firm grasp. Say a country is in need of a petroleum exploration policy in order to make the most effective use of the few small petroleum deposits in its territory. One option would be the usual, science-based one, involving gravity, magnetic, and seismic surveys to pinpoint the most promising locations for drilling. An alternative, initially cheaper option might be to identify promising locations by having the Supreme Leader throw darts at a wall-mounted map of the country.

But as we'll see, the material universe *as a whole* lies outside the domain of science. Thus cosmological schemes—schemes offering descriptions of the birth, evolution, and possible death of the universe—as useful as they are for science—are beyond the competence of science. They are no more and no less valid than analogous descriptions given by religion (one's own) or by myth (the other guy's religion). Because science cannot speak authoritatively about the universe as a whole, we have an opening for the legitimate entry of nonscientific modes of comprehension into the business of explaining the material universe. Here, one's feelings and beliefs can be just as valid as the scientific-appearing descriptions espoused by scientists. They might be even more valid to their holder than those of science, if, for instance, they're more satisfying or aesthetically pleasing.

My own background as a physicist inclines me to cosmological schemes couched in scientific terms. Even so, I do realize the intrinsic lack of scientific validity of such schemes and avoid taking them as seriously as I take genuine science. But more about that later. In any case, if someone prefers the biblical description of the coming into being of the universe, for example, or any other description couched in mythic terms, science cannot object. Even with all its tools and procedures, it really can do no better.

To understand some of the intrinsic limitations of science, in particular its inability to comprehend the material universe as a whole, we must first understand just what science is (and is not). That's what the present chapter and the next are about. The "scientific method," that oversimplification taught us in school, involving observation, hypothesis, experiment, and theory, is only part of the picture. This chapter and the next convey only part of the picture as well, but it is the part I need as groundwork in support of my presentation, and in making my points. So let's get to work, and since we should know what we're talking about, we start with some definitions.

Science

Consider the following definition of science:

> *Science is our attempt to understand objectively the reproducible and predictable aspects of nature.*

Each element of that definition also needs definition. First, look at *our*. This seemingly innocuous qualifier carries a heavy load of implication. It tells us that the source of science lies within ourselves, that science, although having to do with nature, is a human endeavor. Nature would presumably go its merry way, whether we were around or not, or whether we tried to understand it or not. But without our curiosity and our urge to understand, *science* would not exist.

Now for *attempt*. This is our admission that we forgo a priori any claim for assured success. Thus science, in spite of its amazing successes, might not yet have the final word (if indeed there is a final word to be had), and might not yet be capable of handling everything within its domain.

We dealt extensively with *objectively* in the preceding chapter. What we mean is that everything that science does must be as objective as possible. Any evidence used must be objective: available and accessible to everyone who makes the effort. All considerations must be based on reason, which is accessible to everybody, rather than on the likes of intuition, emotion, and faith, which are private.

Next consider *understand*. We take "understand" to mean "be able to explain." That's fine, but what is meant by "explain"? Here we use the dictionary definition: give reasons for. We consider a phenomenon to be understood if we're satisfied that we know the reasons for it.

Nature

Now consider *nature*. For the purpose of our discussion:

> *Nature is the material universe with which we can, or can conceivably, interact.*

The universe is everything. The material universe is everything having a purely material character. To interact with something is to act upon it and be acted upon by it. That implies the possibility of performing observations and measurements on it, and of receiving data from it, which is what we're actually interested in. To be able *conceivably* to interact means that, although we might not be able to interact with that something at present, interaction is not precluded by any principle known to us, and is considered attainable through further effort and further technological advance. Thus the material universe with which we can, or can conceivably, interact is everything of purely material character that we can, or can conceivably, observe and measure. That is what we mean by nature.

"But nature is surely more than that!" many would exclaim. "What about beauty, love, etc.? Aren't they part of nature too?" They, and other such subjective concepts, are certainly part of the universe, but whether they are of purely material character or not remains an open question. So, for the purpose of our presentation, we exclude subjective concepts such as mind, idea, feeling, emotion, and so on, and confine ourselves to the narrow, strictly materialist definition of nature. Thus "nature" serves us as a convenient, concise term for the subject of our present investigation, which is *the material universe with which we can, or can conceivably, interact.* The universe might very well possess other components too, but if so, they're not of concern to us here.

That leaves *reproducible and predictable aspects* to look into, and this too involves some discussion.

Reproducibility

Reproducibility means that experiments can be replicated by the same and other investigators, thus providing data of objective, lasting value about particular aspects of the phenomena of nature. Reproducibility makes science a common human endeavor (rather than, say, an incoherent collection of private, incommensurable efforts). It allows investigators to communicate meaningfully, across space and time, and to progress through joint effort. Reproducibility makes science as nearly as possible an objective endeavor of lasting validity. There seems to be no necessity a priori that nature be reproducible at all, but the fact that science is being done (and redone) proves that nature does possess reproducible aspects. The reproducible aspects of nature form the objective real world that science deals with.

We don't claim that nature is reproducible in all its aspects. But any irreproducible aspects it might be found to possess lie outside the domain of science, by the very definition of science that informs our present investigation. Parapsychological phenomena, for example—extrasensory perception (ESP), telepathy, telekinesis, clairvoyance—if, as some claim, they exist, would nonetheless be irreproducible aspects of nature.

For a detailed view of reproducibility, let's express things in terms of experiments and their results. Reproducibility is then commonly defined by the statement that a given experiment, when replicated, always gives the same result. But what is "the same" experiment? Each experiment, and we're including here all of the runs made on the same experimental apparatus, is a unique phenomenon. We try to avoid all possible changes, but no two experiments are identical. They must differ at least in their calendar and clock times (where the experiment, is repeated in the same laboratory) and/or in their locations (where the experiment is duplicated in another laboratory). And they might (and always do) differ in other aspects as well, such as in their orientations in space. So when we specify "same" experiment and "same" result, we actually mean equivalent rather than identical. We can't even begin to think about reproducibility without permitting ourselves to overlook certain differences, such as those that involve time and location as well as various other aspects of experiments.

Consider the difference between two experiments as being expressed by the change that must be imposed on one experiment if we are to convert it into the other. Such a change might involve a change of time, if the experiments are performed at different times. It might (also) involve a change of location, if they are (also) performed at different locations. If the experimental setups have different orientations in space, the change will involve rotation from one direction to the other. If they are in different states of motion, a change of velocity will be involved. We might replace a brass component of the apparatus with an equivalent plastic one. We might bend the apparatus somewhere. Or we might measure velocity rather than pressure. And so on.

But not all possible changes are associated with reproducibility. Let's list the ones that are. We certainly want to include change of time, to allow the experiment to be repeated in the same laboratory, and rotation and change of location, to allow other laboratories to duplicate the experiment. Since almost all laboratories are attached to Earth, the motion of Earth—a complicated affair compounded of its daily rotation about its axis, its yearly revolution around the Sun, and even the Sun's motion, in which the Earth, along with the whole solar system, participates—requires rotation and changes of location and velocity, both for experiments repeated in the same laboratory and for those duplicated in other laboratories. Then, to allow the use of different sets of apparatus, we need to add replacement by other materials, other atoms, and other elementary particles. Owing to unavoidable limitations on the precision of experiments, we must also include minute changes in the conditions of running the experiments. And there are changes of a more technical nature, which we needn't go into here.

Thus the reproducibility-associated changes that might be imposed on experiments include: change of time, change of direction (rotation), change of location, change of velocity, replacement by other materials, and small changes in the conditions.

As an example, imagine some experiment whose result is a particle appearing at the center of the apparatus one second after we turn on the switch. Repeat the experiment with the same apparatus, in the same direction and state of motion relative to Earth, but eight and a half hours later and at a location 2.2 kilometers east of the original location. If the particle now appears where we expect it eight and a half hours later than, and 2.2 kilometers east of, its previous appearance—that is, if it still appears at the center of the apparatus one second after the switch is thrown—we have evidence that the experiment might be reproducible. (Evidence, but not proof of reproducibility. By the rules of logic, a single negative result disproves reproducibility, but no finite number of positive results can prove it. A few positive results make us suspect reproducibility; many convince us; great numbers of additional positive results confirm our conviction.) In other words, if we ignore when and where the experiment is performed, and we nonetheless obtain the same result every time—a particle appears at the center of the apparatus one second after the experiment starts—then it's looking as if changing time and shifting position don't matter.

So let us define reproducibility. Consider two experiments that differ only by one or more of the reproducibility-associated changes mentioned above (such as change in the time or place the experiment is carried out). For these experiments to be considered *essentially* "the same," even though they are truly different, the *intrinsic* experiment-result process must be identical for both experiments, and thus unaffected by the difference between them. This means the results of the experiments must differ in exactly the same way the experiments differ. In the example, the experiments are 2.2 kilometers apart and so are the results, with the particle appearing at the center of the apparatus in both cases. The experiments are 8.5 hours apart in time, and so are the results: in both cases the particle appears one second after the switch is thrown. The *intrinsic* process—a particle appearing at the center of the apparatus one second into the running of the experiment—is unaffected by the experiments' different locations and times. If this holds true for some experiment, and for its replicas that differ from it by *all* reproducibility-associated changes (that is, all changes of time, direction, location, velocity, and so on), we then declare this experiment reproducible.

The idea is that when any reproducibility-associated change is imposed on a reproducible experiment in its entirety—that is, the same change is imposed on both the experiment and its result—the changed result will always be the actual result of running the changed experiment. One could then say that for such an experiment nature is indifferent to reproducibility-associated changes; they are unessential changes. Impose such a change on a reproducible experiment, and you get an experiment that, although it might differ from the original one in location, orientation, time of execution, and so on, is still *essentially* the same experiment in the sense that its result is *essentially* the same the result.

To put things a different way, imagine some reproducible experiment whose result is an explosion occurring at the edge of the apparatus 2.5 seconds after the switch is turned on. If we repeat the experiment with the same apparatus, in the same direction and state of motion relative to Earth, etc., but three hours later and at a location 1.6 kilometers west of the original location, we're assured the explosion will occur three hours later than and 1.6 kilometers west of its previous occurrence. That is, it will again pop up at the edge of the apparatus 2.5 seconds after the start. And that will happen no matter when or where the experiment is performed. So ignoring when and where, the experiment always gives the same result.

Predictability

Predictability means that among the natural phenomena we choose to investigate, order can be found, from which laws can be formulated, laws that predict the results of relevant new experiments. Predictability makes science a means both to understand nature and to exploit nature. Just as for reproducibility, there seems to be no necessity a priori that nature be predictable at all, but the fact that science is being done, in countless laboratories and other circumstances, proves that nature indeed possesses predictable aspects. And just as for reproducibility, we do not claim that nature is predictable in all its aspects. But any unpredictable aspects it might possess lie outside the domain of science by the definition of science that informs our present investigation. Parapsychological phenomena, for example, if they exist, would be such an unpredictable aspect.

In order to view predictability in detail, let's again express things in terms of experiments and their results. Predicting the results of new experiments doesn't often come about through pure inspiration, but is normally attained by performing experiments, studying their results, finding order among the collected data, and formulating laws that fit the data and consistently predict new results.

Imagine we have an experimental setup, and we run a series of, say, fifty experiments on it. We have experimental inputs $exp_1, \ldots, exp_{50}$ and corresponding experimental results $res_1, \ldots, res_{50}$, respectively. Thus we obtain the fifty data pairs $(exp_1, res_1), \ldots, (exp_{50}, res_{50})$. We study these data, apply experience, insight, and intuition, perhaps plot them in various ways, and, perhaps with a bit of luck, we discover order among them. (There's no standard operating procedure for finding order!)

Suppose we find that all the data pairs obey a certain relation, such that all the results are related to their respective inputs in the same way. This relation is then a candidate for a law predicting the result *res* for *any* experimental input *exp*: simply apply the relation that works so well for the data pairs already obtained to any other input *exp* and thus find the predicted result *res*. Imagine further that this is a valid law. Then additional experiments will confirm it, and we'll find that data pairs (exp_{51}, res_{51}), (exp_{52}, res_{52}), and so on, also obey the same relation, as predicted. Predictability is the existence of such relations for experiments and their results.

For an example, consider the experimental setup of a given sphere rolling down a fixed inclined plane. The experimental procedure consists of releasing the sphere from rest, letting it roll for any time interval t, and noting the distance d the sphere rolls during the time interval. (Here t and d are playing the roles of *exp* and *res*, respectively.) Suppose we perform ten experiments, giving the data pairs $(t_1, d_1), \ldots, (t_{10}, d_{10})$. Or, to be more specific, say we obtain the following ten numerical data pairs, where the time interval t is measured in seconds and the distance d is in meters.

t	0.5	1	1.5	2	2.5	3	3.5	4	4.5	5
d	0.025	0.1	0.225	0.4	0.625	0.9	1.225	1.6	2.025	2.5

A half-second roll from rest covers 0.025 meter. In one second the ball rolls 0.100 meter from rest, in two seconds 0.400 meter from rest, and so on.

We study these data and plot them in various ways. (The example is idealized, since in the real world there are always experimental errors that must be dealt with. But to avoid unduly complicating our presentation here, we ignore such difficulties.) Most of the plots show nothing especially interesting. But lo and behold! In the plot of the distance d against the *square* of the time interval t^2, it looks as if all ten points fall close to a straight line:

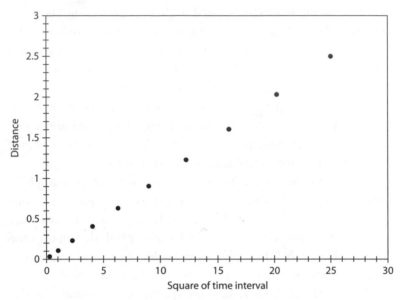

A plot of the data from the rolling-sphere experiment. The distance the sphere rolls from rest is plotted against the square of the time interval, and the data points tend to fall close to a straight line.

The fact we just discovered, that the ten data points all fall near a straight line in the plot of d against t^2, suggests the relation that the distance traveled from rest is proportional to the square of the time interval: $d_1 = bt_1^2, \ldots, d_{10} = bt_{10}^2$, where b is the coefficient of proportionality. This means that when we double the time we allow the ball to roll, for example, the distance it goes is quadrupled. Check that. The distance covered in a two-second roll, 0.4 meter, is four times the distance of 0.1 meter the ball rolls in one second. And similarly for other data pairs.

That suggests the law $d = bt^2$ predicting the distance d for *any* time interval t, not only the ones we measured. For our numerical example, the suggested law is $d = 0.10t^2$. As it happens, this hypothesis is correct, and all additional experiments confirm it: $d_{11} = bt_{11}^2$, $d_{12} = bt_{12}^2$, and so on. Or to be specific, we might run the experiment for the additional time intervals $t = 5.5$ seconds, $t = 6.0$ seconds, and so on. By substituting these values of t in the relation $d = 0.10t^2$, we calculate the respective predicted distances the sphere is predicted to roll for the additional time intervals: $d = 0.10 \times 5.5^2 = 3.025$ meters, $d = 0.10 \times 6.0^2 = 3.60$ meters, and so on. When the experiment is run, these are the actual distances the ball rolls. The new data are found to obey the suggested law. Thus the relation of distance to time interval is a predictable aspect of the setup.

It should now be clear what we mean by *the reproducible and predictable aspects of nature*. They are those aspects of nature that are both objective (all investigators agree on them) and orderly (they exhibit sufficient regularity to allow prediction).

But note this. Since we don't include *the reproducible and predictable aspects* in our definition of nature as *the material universe with which we can, or can conceivably, interact*, it follows that we're leaving the door open to the possible appearance of irreproducible and/or unpredictable phenomena in nature. Such phenomena could be fascinating, and we might, with good reason, attempt to understand them. But then we would not be doing *science*; we would be involved in nonscientific modes of comprehension and explanation, and that would belong to another story altogether.

Law

At this point we should be in a position to begin to appreciate the definition of science that forms the basis of our present investigation:

> *Science is our attempt to understand objectively the reproducible and predictable aspects of nature.*

In our effort to understand, we first search for order among the reproducible phenomena of nature, and than attempt to formulate laws that fit the collected data and predict new results. Such laws of nature are expressions of order, of simplicity. They condense all existing data, as well as any amount of potential data, into compact expressions. Thus, they are abstractions from the sets of data from which they are derived, and are unifying, descriptive devices for their relevant classes of natural phenomena.

In the above example of the rolling sphere, the law $d = bt^2$ is an abstraction from the data pairs (t_1, d_1), (t_2, d_2), . . . , thus a simplification of them. It expresses an order that exists among the different runs of the experiment. It offers a description and a unification of the rolls of the sphere down the plane.

As an archetypal historical example of a law of nature derived from observational data, let's consider the case of Kepler and his three laws of planetary motion. Johannes Kepler (1571–1630) pondered the astronomical data available to him for the five planets that were known at his time—Mercury, Venus, Mars, Jupiter, and Saturn—and found this order. If Earth is considered as one of the planets, and if the motions of the planets (the six innermost planets of the solar system) are considered from the point of view of a hypothetical observer standing on the Sun (the heliocentric point of view), then the motions possess three properties. For our present discussion it's not necessary to understand the properties in detail, and if you're unfamiliar with the concepts involved, don't worry. The properties are:

1. *The path each planet traverses in space, its orbit, lies wholly in a fixed plane and has the form of an ellipse, of which the Sun is located at a focus.* (Actually, the ellipses of these six planets are close to being circles, with the Sun located at their common center.)

2. *As each planet moves along its elliptical orbit, the (imaginary) line connecting it with the Sun sweeps out equal areas during equal time intervals.* These areas have the shapes of slices from an elliptical pizza. (Thus, from geometric considerations, a planet moves faster when it's closer to the Sun and more slowly when it's farther from

the Sun. Since the orbits of these six planets are nearly circular, this property simply means that each planet moves with nearly constant speed.)

3. *The ratio of the squares of the orbital periods of any two planets equals the ratio of the cubes of their respective orbital major axes.* The orbital period of a planet is the time it takes to complete one revolution around the Sun, the planet's "year." The orbital major axis of a planet is the major axis of the ellipse formed by its orbit, which is the distance between the two ends of the ellipse, one end being the point of nearest approach to the Sun and the other the point of farthest departure from the Sun. (Since the orbits of the six planets are nearly circular, the major axis is practically the diameter of the orbit, so without much error one may read "orbital diameters" for "orbital major axes." And since those lengths appear in a ratio so that a power of two cancels, one may also read "orbital radii.") If we denote the orbital period of any one of the planets by T_1 and that of any other by T_2 and denote their respective orbital major axes (or diameters or radii of their orbits) by a_1 and a_2, then this property is expressed by the formula

$$\frac{T_1^2}{T_2^2} = \frac{a_1^{\,3}}{a_2^{\,3}},$$

or equivalently

$$\left(\frac{T_1}{T_2}\right)^2 = \left(\frac{a_1}{a_2}\right)^3.$$

As I mentioned just before presenting the three properties, it's not necessary that you understand the properties in detail. The most important point is that Kepler formulated properties of the planets' motions. It is of secondary importance that the properties concern (1) the form of the planetary orbits, (2) the speed of each planet as it moves along its orbit, and (3) the relation between the time it takes each planet to complete one revolution around the Sun (the planet's "year") and the size of its orbit.

These properties express an order among the astronomical data. They offer a description and a unification of them. The motions of the six planets are not just any motions, but are related by their possessing the three properties.

The properties are called Kepler's laws of planetary motion. They are laws of nature, in the sense that they correctly predicted the relevant properties of the motions of the additional planets that were discovered in the solar system in later years: Uranus, Neptune, and Pluto (no longer considered to be in the planet category). They are also laws in the sense that they are valid for any system of astronomical bodies revolving around a massive central body, such as the moons revolving around the planet Jupiter.

So now we have science proceeding by our (1) collecting reproducible, thus objective, data about nature and (2) finding, by whatever means, objective order and laws among the data. But laws are not explanations; they are descriptions. They describe the data in a concise manner and allow predictions. But they do not explain why things are as they are. To *understand* objectively the reproducible and predictable aspects of nature and not merely describe them, we must take an additional step. We must find objective explanations for the order and laws we discover. Such an explanation is called a theory. The next chapter deals with theories.

ℬ

Science is our attempt to understand objectively the reproducible and predictable aspects of nature, where nature is taken to mean the material universe with which we can, or can conceivably, interact. The conduct of science involves searching for order among the reproducible phenomena of nature and then attempting to formulate laws that describe the collected data and predict new results.

Bibliography

For an idea of what physics, my model of a natural science, is about, see
R. K. Adair, *The Great Design: Particles, Fields, and Creation* (Oxford University Press, Oxford, 1987).

R. G. Newton, *The Truth of Science: Physical Theories and Reality* (Harvard University Press, Cambridge, Mass., 1997).

———. *What Makes Nature Tick?* (Harvard University Press, Cambridge, Mass., 1993).

J. S. Trefil, *Reading the Mind of God: In Search of the Principle of Universality* (Charles Scribner's Sons, New York, 1989).

For more reading about the nature of science, see

H. Fritzsch, *The Creation of Matter: The Universe from Beginning to End* (Basic Books, New York, 1984).

R. Morris, *Dismantling the Universe: The Nature of Scientific Discovery* (Simon and Schuster, New York, 1983).

I. Prigogine and I. Stengers, *Order out of Chaos: Man's New Dialogue with Nature* (Bantam Books, New York, 1984).

For the purposes of the present book, it's not necessary to understand Kepler's laws or even to be familiar with them. But if you're interested, I suggest you look at any introductory college physics textbook. For a historical perspective of the laws, see

D. Park, *The How and the Why* (Princeton University Press, Princeton, N.J., 1988).

For the significance of the laws of nature, and for symmetry as an expression of the simplicity of those laws, see

R. P. Feynman, *The Character of Physical Law* (MIT Press, Cambridge, Mass., 1965).

See also Newton (1993) and Park (1988), above.

For introductory presentations of symmetry, see

G. Darvas, *Symmetry* (Birkhäuser, Basel, 2007).

J. Rosen, *Symmetry Discovered: Concepts and Applications in Nature and Science* (Cambridge University Press, Cambridge, 1975; reprinted with additions by Dover Publications, Mineola, N.Y., 1998).

More on laws of nature can be found in

J. D. Barrow, *The World within the World* (Oxford University Press, Oxford, 1988).

H. R. Pagels, *The Cosmic Code: Quantum Physics as the Language of Nature* (Simon and Schuster, New York, 1982).

V. J. Stenger, *The Comprehensible Cosmos: Where Do the Laws of Physics Come from?* (Prometheus, Amherst, N.Y., 2006).

Some books on irreproducible and unpredictable phenomena are

H. Broch, *Exposed! Ouija, Firewalking, and Other Gibberish* (Johns Hopkins University Press, Baltimore, 2009).

G. Charpak and H. Broch, *Debunked! ESP, Telekinesis, and Other Pseudoscience* (Johns Hopkins University Press, Baltimore, 2004).

M. Shermer, *Why People Believe Weird Things: Pseudoscience, Superstition, and Other Confusions of Our Time* (Freeman, San Francisco, 1997).

CHAPTER THREE

Theory

Explanation, Not Speculation

Theory

LAWS OF NATURE are worthwhile achievements. Besides their potentially useful predictive power, they offer a unifying description of natural phenomena. But we are not satisfied with that. We want to *explain* laws of nature, to know the reasons for them; we want to *understand* the reproducible and predictable aspects of nature, not just describe and predict their phenomena. And we want to do it in an *objective* way, in a manner valid for all. That is science in the big picture. Scientists' term for an explanation of a law of nature is *theory*.

There are a number of criteria by which theories are judged for their acceptance or rejection in science. These criteria are not intrinsic to nature itself, are not imposed *on us* by nature in any way, but are imposed *by us* in our search for understanding. The criteria, however, are but rationalization. What determines a theory's acceptability is its giving us the *feeling* that something is indeed being explained, so that it satisfies our curiosity about the reasons for whatever law we're trying to understand. The detailed criteria are our attempt to rationalize that feeling. Yet it is difficult, if not impossible, to meaningfully communicate such feelings, and we have no choice but to discuss the criteria.

Feeling? Satisfy? What's going on here? Are we discussing science or not? Aren't those terms associated with subjective, nonscientific modes of comprehension? Yes, they are. And in that respect they're connected with the irrational side of human behavior. Since, as we saw in the section *Science* of the preceding chapter, science is a human endeavor, all aspects of human behavior can manifest themselves in the doing of science, even irrationality. We try to rationalize our irrationality, of course. But it is unavoidably there, just as in any other human activity. We'll see more of it farther on.

Yet in spite of that, the scientific mode of comprehension is still the most nearly objective mode of comprehension around, and we'll have to make do with it. Actually, we make do very well, thank you. Nevertheless, an open-eyed realization of the irrational aspects of science as a human activity cannot fail to benefit our attempt to clarify the role of science in our understanding of the world around us.

Logical Implication and Objective Truth

One of the most important properties of an acceptable theory, part of its *prerequisite*, is that whatever is doing the explaining must, within the logical framework of the theory, actually imply what is being explained. In other words, as was pointed out in *Logical Truth*, chapter 1, an acceptable theory must be logically true. A "theory" that does not satisfy this criterion cannot be considered to be an explanation at all, cannot be seen as giving reasons for anything. That should be obvious. Still, a fanciful example of such a "theory" could be this. Planets are similar to apples: both are round. There are varieties of apples that are red. The word "red" is spelled with three letters. Kepler's laws of planetary motion are also three in number. Thus Kepler's laws must hold.

As an aside, let me point out that, as artificial and absurd as this "theory" appears to be, in pre-enlightenment times mystical and numerological considerations and justifications were perfectly acceptable. The theories of those times read like fantasies to (most of) us today.

Logical implication reaches farther than it might appear. If a theory logically implies more than it was originally intended to explain, then it predicts new laws. But these new laws must be found to be valid themselves, otherwise, the parent theory is false as well. Thus, related to the property of logical implication, and another part of the prerequisite, is the other most important property of an acceptable theory, the property of being true in the sense that nothing implied by a theory should contradict experimental findings. In chapter 1 we called that being objectively true. (This property is also related to the property of falsifiability, discussed later in this chapter.)

As an example, assume we have a theory to explain Kepler's laws of planetary motion, one that is also found to predict laws of motion for the comets, although cometary motion was not taken into account when the theory was devised. If the actual motions of the comets do not obey the laws predicted by the theory, then the theory is deemed objectively false and thus unacceptable, in spite of its apparent success in explaining planetary motion.

Generality and Fundamentality

Two additional properties of an acceptable theory, not prerequisite but still very important, are that what is doing the explaining should be *more general* than what is being explained and should also be *more fundamental*. *Generality* is usually easy enough to discern: the more general the category, the larger the number of natural phenomena it encompasses. Theories found to possess the property of explaining more than they were originally intended to explain, thus predicting new laws, are said to exhibit generality. Possessing that property allows the theory to be tested by comparing the predictions of these laws with the results of experiment. We made mention of this in the preceding section: the property of a theory that it can be tested in this way is called *falsifiability*.

For example, a theory of Kepler's laws based on the motion of the Earth would not be acceptable to scientists, since the motion of a single planet is less general than the motions of all the planets. On the other hand, a theory of Kepler's laws based on universal laws of motion for all bodies could be acceptable, because laws of motion for all bodies are more general than laws of motion valid only for planets. Such a theory might then predict laws of motion for, say, comets, and those laws could then be compared with the actual motions of comets.

Fundamentality is less simple than generality, since it depends on one's scientific worldview and on the context of discourse. What you take to be fundamental might seem derivative to me, while what is underlying for me might appear superficial to you. Scientists working within the commonly accepted scientific worldview tend to agree on questions of fundamentality. Still, disagreements can arise. (Note the subjectivity, thus irrationality, here.)

As an example, a theory of genetics based on some macroscopic property of an organism, such as body weight, would not be acceptable to scientists, because heredity is generally considered to be more fundamental than, and to determine to a certain extent, the macroscopic and microscopic properties of organisms. But since biochemistry is commonly considered to underlie biological phenomena, a molecular theory of genetics, such as the currently successful one involving DNA, RNA, etc., could be acceptable.

Here is an example of disagreement over fundamentality. There is a class of theories that attempt to explain aspects of the universe as a whole by the existence of human beings. These theories are unacceptable to most scientists, since the existence of human beings is generally considered to be much less fundamental than anything having to do with the universe as a whole. Proponents of these theories claim, however, that in a certain sense our existence *can* be conceived as being more fundamental than the universe as a whole. This approach, called the *anthropic principle*, is discussed in detail in chapter 7.

Naturality

Another important property of a scientifically acceptable theory is that what is doing the explaining should itself possess a natural character. In other words, a theory should explain one aspect of nature by another, and should not look outside nature for its explanations. (Recall that nature has the specific and narrow meaning we assigned it in *Nature*, chapter 2.) For example, "The apple falls by the will of God" cannot be a scientific theory, since godhood, by definition, lies outside nature. And thus also do creationism and "intelligent design" fail as scientific theories. But "Our fate is in the stars," if it were true, could possibly be an acceptable theory.

The criterion of naturality is not always strictly adhered to, however, and there are controversies among scientists about just how loose a theory can be in this regard and still be acceptable as scientific. The problem has to do with theories concerning aspects of nature that are generally considered to be among the most fundamental of all, such as space and time, the properties of the elementary particles, and the evolution of the universe. Put simply, to explain a most fundamental aspect of nature by something even more fundamental, one is forced to go beyond nature.

There are contemporary theories that are based on inherently undetectable extra dimensions, that involve other universes with which we cannot conceivably interact, or that consider the situation prevailing "before" the coming into being of the universe and thus "outside" space and time themselves, for example. Some scientists consider such theories unacceptable. They feel that no explaining is being done. But whether those ideas are accepted as explanations or not, they certainly do offer unifying frameworks, tying together diverse aspects of nature such as the properties of the elementary particles and the evolution and large-scale properties of the universe. So we see that in such a pinch some scientists tend to prefer to stick with fundamentality even at the expense of naturality.

One might wonder why something like "The apple falls by the will of God" is unacceptable as a scientific theory, while an explanation involving undetectable extra dimensions or other universes can be embraced by scientists. Both God and extra dimensions lie outside nature, so neither kind of theory possesses naturality. What is the difference? The answer is that God-type theories possess few or none of the other desirable or essential properties of scientific theories, neither those properties we have discussed above (logical truth, objective truth, generality, fundamentality) nor those to be discussed in the following (causation, simplicity, unification, beauty, falsifiability). Proposed theories of the extra-dimensions type possess most or all of those properties. That is the difference, a significant difference.

Causation

Another advantageous property of an acceptable theory is that what is doing the explaining should be perceived, not merely as logically *implying* what is being explained, but as *causing* it. This means that some sort of causal linkage, some "mechanism," should be perceived as joining what is doing the explaining (the cause) with what is being explained (the effect). It is the *perception* of such a causal linkage that enhances the acceptability of a theory. Whether there "actually exists" a causal mechanism or not can depend on one's point of view, even on one's scientific worldview, and so is not an objective property of nature. (Again note the subjectivity here.)

For example, a theory explaining Kepler's laws might consist of a statement of general laws of motion valid for all bodies, not just for planets. Kepler's laws, describing the motion of the solar system as a special case, would be derivable by mathematical means from those general laws. Thus, although the theory would have the property of logical (mathematical) implication, it would most likely not arouse the perception of causation, so that some might find such a theory not completely satisfying.

At the same time, a theory of Kepler's laws might be stated, from a somewhat different point of view, as general laws of motion together with the existence of the Sun and the planets' attraction to the Sun. Such a theory would most likely arouse the perception that the Sun *causes* the planets' special motion through its pull on them. As a matter of fact, the standard theory of Kepler's laws is just of this kind, as we will soon see.

Simplicity and Unification

Two other favorable properties of an acceptable theory are that what is doing the explaining should be simpler than what is being explained, and should also be more unifying than the latter. And the simpler and more unifying it is, the more acceptable the theory. Although some criteria for simplicity can be stated (nevertheless, we will not), a generally satisfactory objectification of the concept has not been achieved. Simplicity is largely a matter of subjective perception. (Yes, subjectivity again.) Simplicity depends very much on one's taste and worldview, even on one's education, although there does seem to exist a considerable degree of consensus about it among scientists working in the same field, perhaps not surprisingly. In any case, scientists prefer "simple" theories to "complicated" ones, however simplicity is judged.

For example, Albert Einstein's (1879–1955) general theory of relativity is one of a number of theories of gravitation. (Let me remind you that gravitation is the universal force of attraction between all pairs of bodies in the universe, including the force holding us firmly down to Earth so that we don't float off into space.) All these theories appear to be overwhelmingly complicated—even to most scientists. Still, theoretical physicists specializing in the field generally perceive Einstein's theory as the simplest, and thus the preferred, one.

Unification is easier than simplicity to pin down. As a general rule, the more numerous the different concepts a theory involves, the more unifying it is. Thus a theory should tie together and interrelate more aspects of nature than are tied together and interrelated by what is being explained by the theory. For example, the concepts involved in Kepler's laws are purely kinematic; they are solely concepts of motion: position, orbit, area, time, speed, and so on. Any explanation involving only these same concepts would not be more unifying than Kepler's laws themselves. But a theory involving kinematic concepts along with additional ones, such as force and mass, would be more unifying. Unification can generally be expected to go hand in hand with generality.

Beauty

We now come to a matter that might seem absolutely amazing, which is beauty in theories. To many, science has the image of a cold, rational endeavor, to which considerations such as aesthetics are perfect strangers. Yet, although rationality is indeed the major ingredient of science, aesthetic considerations are far from foreign to it. (I warned earlier in this chapter to expect irrationality in science!) Scientists are constantly heard referring to "beautiful ideas," "beautiful experiments," "beautiful laws," and "beautiful theories." And a scientist will always prefer a beautiful theory to an ugly one, other things being equal, and even often at the expense of some other desirable property of acceptable theories. Many scientists admit that the pleasure they derive from their profession has a large aesthetic component, and for some (including myself) the aesthetic consideration is predominant.

What is a beautiful theory? Well, here we go again. Just as an acceptable theory is one that gives the feeling that something is being explained, so a beautiful theory is one that arouses the feeling of beauty. Beauty is not an objective property of theories, but is wholly in the eyes of the beholder and completely subjective. So the most we can do is to rationalize again and try to point out those properties of theories that contribute to their perceived beauty. It appears to me that the main contribution to the beauty of a theory derives from its simplicity, its unification, and its generality. A theory deemed beautiful by scientists working in the relevant field is invariably simple, greatly unifying, and of broad generality.

Scientists' preference for beautiful theories is truly astounding, when we consider that beauty is subjective and that there *seems* to be nothing in nature that *requires* beauty. This preference finds expression in scientists' irrational and objectively unfounded conviction that nature *must* be understandable in terms of beautiful theories. The alternative is simply inconceivable. Even more astounding is the fact that successful theories do tend to be beautiful! (Or perhaps that's not so astounding after all, considering that theories are part of science and science is a human endeavor. More about that in chapter 7.)

In this connection, I relate the story of Paul Adrien Maurice Dirac (1902–1984), who developed a beautiful theory of the electron. (The electron is a type of elementary particle, one of the constituents of atoms, and is, by its relatively free motion in metals, the carrier of electric current in cables and wires.) Dirac's theory predicted the existence of another type of elementary particle, very different from the electron in at least one respect (having the opposite sign of electric charge), yet identical to the electron in other respects (such as mass and magnitude of electric charge).

At that time the only other types of elementary particle known were the proton and the neutron (both are constituents of the nuclei of atoms), and neither of them satisfied the specifications of Dirac's predicted particle. Thus the theory was considered objectively false. But Dirac didn't relinquish his theory, and the eventual discovery of the positron proved him right. The moral, according to Dirac, is that "it is more important to have beauty in one's equations than to have them fit experiment."

One who didn't follow Dirac's policy and lived to regret it was Erwin Schrödinger (1887–1961). Schrödinger devised a beautiful theory, involving a certain equation, to explain atomic phenomena. But when he applied it to the hydrogen atom, the simplest atomic system, he obtained results that were in disagreement with experiment. Then he noticed that a rough approximation to his equation gave results that did agree with experimental observations. So he published his approximate theory, a much less beautiful one than the original. Because of his delay, the original theory was published by others and credited to them. What happened was that the original, beautiful theory was not appropriate to the electron in the hydrogen atom, but was suitable to types of elementary particle that had not yet been discovered. The approximate, uglier theory was insensitive to the differences between the types of particle and was fairly accurate when applied to the hydrogen atom.

Falsifiability

Falsifiability amounts to testability, the property that a theory can be tested against as-yet-unknown natural phenomena to determine whether it is objectively true or false. This property, not directly related to a theory's ability to give the feeling that something is being explained, is generally required of any acceptable theory. Karl Popper (1902–1994) introduced the idea in the 1930s as the criterion one should use to distinguish science from pseudoscience. To be falsifiable, a theory must predict something in addition to what it was originally intended to explain (and which it presumably does explain; otherwise it would not be a candidate for an acceptable theory). Its prediction is tested against experimental results that were not taken into account when the theory was devised (either because they had not yet been obtained or because they were not known to the deviser of the theory). To produce predictions, a theory must fulfill the criterion of generality, as we saw above. Then what is doing the explaining, being more general than what is being explained, can explain more than it was originally intended to explain and thus make testable predictions.

A falsifiable theory is in constant danger of being invalidated by even a single new experimental result. An *un*falsifiable theory, by contrast is, by its unfalsifiability, immune to invalidation. As an example of an unfalsifiable theory, imagine that a theory of the electron is found that explains all the known properties of the electron, but, unlike Dirac's theory, predicts absolutely nothing in addition to that. Such a theory cannot be tested. Even if new properties of the electron are eventually discovered, the theory will not be invalidated thereby. It will still be valid, since it continues, correctly, to explain what has become only part of the electron's properties, whereas it has nothing at all to say about the other properties, and, moreover, does not predict their nonexistence. As a putative theory of the electron, it is incomplete. A falsifiable theory is discussed in the following example.

An Archetypal Example

An archetypal example of a theory is the explanation of Kepler's laws of planetary motion by Newton's laws of motion and gravitation. Newton pondered Kepler's laws of planetary motion (presented in the section *Law* in the preceding chapter) and succeeded in formulating a number of laws of his own in order to explain them. Newton's laws were based on observational data, as were Kepler's, but were also based on Kepler's laws themselves as input. Newton's laws are somewhat counterintuitive, and their full experimental verification had to wait for future technological developments and refinements.

Newton proposed three universal laws of motion, whose modern formulation is more or less this:

1. *In the absence of forces acting on it, or when such forces cancel each other, a body will remain at rest or continue to move uniformly in a straight line.*
2. *A force acting on a body will cause the body to undergo acceleration whose direction is that of the force and whose magnitude is proportional to that of the force divided by the body's mass.* Acceleration is change of velocity—that is, change of speed and/or direction of motion—per unit time. Mass measures the amount of matter in a body. (It is related to, though essentially different from, a body's weight.)
3. *For every force acting on it, a body will react upon the force's source with a force of opposite direction and equal magnitude along the same line of action.*

To these Newton added the universal law of gravitation: *Every pair of bodies undergoes mutual attraction, with the force acting on each body proportional to the product of the bodies' masses and inversely proportional to the square of their separation.*

For our present purpose it's not necessary to understand Newton's laws in detail, and please don't worry if the concepts involved are unfamiliar. The important point is that Newton formulated universal laws of motion and a universal law of gravitation, valid for all bodies in the universe.

These four laws form Newton's theory to explain Kepler's laws of planetary motion (as well as a vast realm of other mechanical phenomena). How are they a theory? First of all, they do logically imply Kepler's laws. With the appropriate mathematical tools, scientists and science students can show that the motions of bodies around a massive body will, under certain conditions, obey Kepler's laws. These conditions are fulfilled by the planets in their motion around the Sun as well as by additional astronomical systems, such as the moons of the planet Jupiter in their motion around Jupiter.

Newton's laws are certainly more general than Kepler's, in the sense that they hold for all bodies (hence the qualifier "universal") and not just for planetary systems. As a result, Newton's laws explain far more than Kepler's; they explain a wealth of mechanical phenomena, earthbound (such as falling apples) as well as astronomical. In addition, they are more unifying, since they show order among broader classes of phenomena than do Kepler's laws. Whereas the latter involve solely concepts of motion, Newton's laws involve also the concepts of force and mass. And they are also more unifying, in that, for example, the motions of comets are shown by Newton's laws to be akin to the motions of the planets, whereas Kepler's laws ignore cometary motion.

Whether Newton's laws are more fundamental and simpler than Kepler's is a matter of one's worldview and taste concerning fundamentality and simplicity, but scientists generally consider them as such. We won't go into details here, but let's just note in connection with fundamentality that, whereas Kepler's laws merely describe motion, Newton's laws have to do with the *causes* of motion, that is, with forces (and their absence). The character of Newton's laws is just as natural as that of Kepler's, and so we have one aspect of nature being explained by another.

As they are stated, Newton's laws might not immediately arouse the perception of causation. However, they imply that the motion of every body is determined by all other bodies through its being gravitationally attracted to them (and possibly also affected by them through additional kinds of force). Thus all other bodies are perceived as the cause of a body's motion by the mechanism of gravitational attraction (and possibly other forces). Newton's laws give scientists the feeling that Kepler's laws are being explained; they satisfy scientists' curiosity about the reasons for Kepler's laws to hold. In addition to all that, Newton's laws possess the property of falsifiability, since they predict so much more than they were originally intended to explain.

𝒜

In the conduct of science, after finding laws of nature, we try to understand, that is, explain, objectively those laws by means of theories. For a theory to be acceptable, whatever is doing the explaining must logically imply what is being explained. The former should form an aspect of nature just as much as the latter should. In addition, the former should be more general, more fundamental, more unifying, and simpler than the latter, and should be perceived as causing the latter. Beautiful theories are preferred. Theories should be falsifiable.

Bibliography

For the purpose of our discussions, it's not necessary to understand Newton's laws or even to be familiar with them. Still, if you are interested, I suggest you see any introductory college physics textbook.

Many of the bibliography entries of the preceding chapter are relevant to this chapter too. Independently of that, here are a few suggestions for specific subjects.

For a historical perspective on Newton's laws, see

D. Park, *The How and the Why* (Princeton University Press, Princeton, N.J., 1988).

For theory, see

F. Rohrlich, *From Paradox to Reality: Our New Concepts of the Physical World* (Cambridge University Press, Cambridge, 1987).

For beauty in science, specifically in physics, see

R. G. Newton, *What Makes Nature Tick?* (Harvard University Press, Cambridge, Mass., 1993).

A. Zee, *Fearful Symmetry: The Search for Beauty in Modern Physics* (Macmillan, New York, 1986, and Princeton University Press, Princeton, N.J., 2007).

For introductory presentations of symmetry, see

G. Darvas, *Symmetry* (Birkhäuser, Basel, 2007).

J. Rosen, *Symmetry Discovered: Concepts and Applications in Nature and Science* (Cambridge University Press, Cambridge, 1975; reprinted with additions by Dover Publications, Mineola, N.Y., 1998).

CHAPTER FOUR

Is Science the *Whole* Story?

Science and Metaphysics

THE DEFINITION OF *science* presented earlier (*Science*, chapter 2) and forming the basis of our present investigation is *our attempt to understand objectively the reproducible and predictable aspects of nature.* The definition of *nature* is strictly limited to *the material universe with which we can, or can conceivably, interact,* which is what our investigation is concerned with (*Nature*, chapter 2). Yet in the same chapter and in chapter 3 we mentioned going beyond nature, and we referred to worldview, taste, fundamentality, simplicity, perceived causation, and beauty, all of these subjective and lying outside the strict limits of nature, thus beyond the domain of science. That brings us to *metaphysics.*

Metaphysics is a branch of philosophy dealing with being and reality. In this book, we use the term *metaphysics* in its narrower sense of the philosophical framework in which science operates. In this sense, metaphysics is concerned with what lies around, below, above, before, and beyond science. For instance, whereas science involves observation of nature, metaphysics might consider the significance of the observer-observed dichotomy (since observers, after all, are also part of nature). Or, whereas science searches for and finds order and laws of nature, metaphysics might consider what constitutes evincive evidence, convincing confirmation, and persuasive proof, and why there *are* order and laws of nature at all. Or again, whereas science is our attempt to understand objectively the reproducible and predictable aspects of the material universe with which we can, or can conceivably, interact, metaphysics might consider modes of existence other than material existence, aspects of the universe with which we cannot, even conceivably, interact, or irreproducible or unpredictable aspects of the universe. Thus, concepts such as worldview, taste, fundamentality, simplicity, perceived causation, and beauty, which lie beyond the domain of science, belong to the domain of metaphysics.

Considerations of science and of metaphysics must not be confused with each other. When we do science, we should stick to scientific considerations, avoiding metaphysical ones, for as wide a range of phenomena as possible, but remaining within the domain of science. And when we choose to do metaphysics, we must be sure the subject of our investigation lies beyond the domain of science.

Science is not concerned with phenomena or concepts that are not aspects of the material universe (such as gods) or that are aspects of the material universe that we cannot, or cannot conceivably, interact with (such as other island universes, described in the following chapter under *The Lawless Universe*). Neither is science concerned with phenomena or concepts that are unpredictable (such as ESP) or are irreproducible (such as ESP again). If such a phenomenon, concept, etc. is related to the philosophic framework in which we do science, it is grist for the mill of metaphysics. So theories, for example, are scientific stuff, but their possible beauty is a metaphysical matter.

A worldview is one's (subjective) attitude toward and interpretation of reality; it is the conceptual framework within which one organizes one's perceptions. Worldviews lie well within the domain of metaphysics, according to our usage, since they can strongly influence the way science is done. As was mentioned in the preceding chapter, issues of taste, fundamentality, simplicity, perceived causation, and beauty in science are subject to one's worldview, and different people may judge them differently.

In science, there are rather strict criteria for truth. Whatever one might think, or whatever controversy might arise concerning nature, in the final analysis experiments are performed, observations are made, and it is nature itself that is the supreme arbiter. I would be remiss here if I didn't warn you that the matter is not as clear-cut as it might appear from the preceding two sentences. There are questions and controversies, mostly of a metaphysical character, about just what objective truth means (if anything), how it might be attained through science (if it is attainable at all), and so on. But none of that is important for our discussion, nor will it detract from the points I'll make. On the whole, scientists have no problems with such issues.

The situation is not the same in metaphysics, including worldviews. Metaphysical considerations are unimpeded by any burden of obligation to be objectively true, although they might be expected not to contradict objective truths, and to be logically self-consistent, that is, not to contradict themselves. At least, I would expect that. But even those expectations have no substantial foundation, and one can find metaphysical positions that do contradict the objective truth of science, or that are logically inconsistent, or both.

A well-known example of that is the biblical description of how Earth came to be populated by animals. It is described as a unique, divinely caused event occurring within a single day. A literal belief in this description, commonly referred to as creationism, stands in stark contradiction to the scientific description of the phenomenon as a natural evolutionary process taking many millions of years.

Thus personal taste and preference reign supreme in metaphysics, whereas in science their influence, though present, is not of overwhelming importance. In connection with that, compare the abundance, intensity, and acrimony of controversy, sometimes developing into personal animosity, in philosophy circles with the general moderation and civility of controversy among scientists. (That, at any rate, is my impression of the situation.)

Transcendence and Nontranscendence

Let's compare two types of worldview: transcendent and nontranscendent. The former are worldviews involving the existence of a reality beyond, or transcending, nature. Nature, with which science is concerned, is viewed as being embedded in, being part of, that transcendent reality. Religions and worldviews involving supreme beings and creators are of this type, although transcendent worldviews are not necessarily religions, nor do they have to involve supreme beings or creators. For example, one might hold the teleological belief that nature operates toward some end, for the accomplishment of some purpose (which may or may not be known to the believer), without necessarily also believing in the existence of a supreme power bringing that about. Or, one might believe the universe had a beginning, that it came into being in some sense, without necessarily also believing in a creator.

In this connection, there is no contradiction in scientists' holding transcendent worldviews. There are religious scientists. (I recall seeing a recent statistic, however, that, at least in the United States, the percentage of scientists who claim to be religious is far smaller than the percentage of the general public claiming to be religious.) The essential point is that in their scientific work they confine their considerations solely to nature. Thus "creation science" is an oxymoron. If one's "science" is directed toward supporting one's creationist, or other religious or transcendent, worldview, then one is not doing true science.

Nontranscendent worldviews, for their part, eschew any reality transcending nature, and make do with considering nature to be all there is. Although denying to deities, creators, and supreme beings any reality, the nontranscendentist does not have to deny the existence of mind, consciousness, thought, emotion, feeling, etc. For the nontranscendentist, such concepts are aspects of the material universe, specifically as activity going on in that extremely complex organ, the brain. The undeniable fact that we are at present far from understanding how mind, consciousness, and the like are realized by brain activity is no deterrence to holding a nontranscendent worldview. The human brain is, after all and as far as I know, the most complex compact system in the universe. Even after eliminating some of the hype, the human brain, and even the mammalian brain, is indeed unimaginably complex.

The usual objection to that is the following. Brain cells are not conscious. Electric currents do not feel. Neurotransmitters do not think. Electrochemical reactions do not emote. So how can mind, consciousness, feeling, and so on be realized by brain activity, which is nothing but the activity of brain cells, neurotransmitters, electric currents, electrochemical reactions, and other such mindless, unfeeling material components? However—and this is the crucial point—complex systems can possess properties that are completely irrelevant and meaningless for their component parts. Specifically, brain activity might give rise in a natural manner to mind, consciousness, feeling, etc., in spite of the fact that the components of this activity—brain cells, electric currents, neurotransmitters, and so on—do not possess mind, consciousness, and the like. Thought, feeling, etc. might be realizations of the integrated activity of the brain *as a whole*, although they cannot be attributed to any component of that activity.

A well-known example of a complex system possessing properties that are irrelevant and meaningless for its component parts is this. Consider a few molecules of gas, such as air, confined in a container. The state of these molecules is fully described by their locations and velocities at any time. As time goes on, the molecules fly about, rebounding from the walls of the container and occasionally, albeit rarely, colliding with each other. Now imagine increasing the number of gas molecules in the container immensely, to many thousands of millions and more. Whatever the number of molecules, their state is still describable, at least in principle, by their locations and velocities at any given time, although it might not be practical to describe the situation in this way because of the tremendous amount of data involved. And the molecules still fly about, rebounding from the walls of the container, and as their number is increased, colliding with each other more and more frequently.

But as the number of gas molecules in the container is increased, the situation can better and better be described in macroscopic terms, in terms of volume of gas, its pressure on the container walls, and its temperature. The macroscopic properties of volume, pressure, and temperature are relevant and meaningful for an ensemble of very many molecules, whereas they are irrelevant and meaningless for each of the individual molecules the ensemble comprises.

Another well-known example of a complex system possessing properties that are irrelevant and meaningless for its components is an insect colony, such as a colony of ants or bees. An insect colony, aptly described as a higher-level, or super, organism, possesses the properties of will for survival, adaptation to weather, foresight (preparation for difficult seasons), self-perpetuation (by reproduction), and so on. The behavior of each member of the colony, however, is characterized by automatic reaction to a limited range of tactile, chemical, auditory, and visual stimuli, and an individual member of one of these colonies, when separated from its colony, does not last long.

None of the macroscopic properties of the colony is possessed by any of its members. A soldier will easily get itself killed in defensive or aggressive action of the colony as a whole. A worker might wear itself out bringing food for the colony. An individual might drown while protecting the colony from rain. Only the queen and a drone reproduce at all, and they reproduce the whole colony. The food collected for the winter is intended primarily for the queen and her offspring, so in case of winter shortage a worker might not get to eat any of the food it brought in. And so on.

Life itself, generally considered by transcendentists to belong to the extranatural domain (as spirit, soul, etc.), is viewed by nontranscendentists as a property that can develop naturally in sufficiently complex systems, in analogy with the above examples, although this is still far from being understood. For each of the individual atoms making up a living organism, the property of life is irrelevant and meaningless; atoms are not alive. Indeed, throughout its lifetime an organism is constantly exchanging atoms with its environment, and any particular atom might form part of an organism for a limited time only.

The nontranscendentist view of mind, consciousness, etc., is similar. They are considered to be properties that are developed naturally by sufficiently complex systems, not even necessarily living systems. These properties are irrelevant and meaningless for each of the individual neurons comprising the brain; cells are not considered capable of thought or feeling. According to this physicalist point of view, sufficiently complex computers might eventually attain the properties of mind or consciousness. We can't know what such computers will be made of, but for whatever will fulfill the function of today's transistor, those properties will be irrelevant and meaningless; individual logic elements do not think or feel.

In this connection, some readers might remember the computer Hal in the classic 1968 film *2001: A Space Odyssey*. Hal is the onboard operating computer of a spacecraft. It (I almost wrote "he") exhibits consciousness and feeling, and usurps control of the spacecraft from the humans it is supposed to obey. Here we are, well past the year 2001, and our best computers are nowhere nearly as capable as Hal. (If you haven't seen the film, or even if you have, I suggest you do. It's a good show.)

Although it might eventually be possible to determine whether life, mind, consciousness, and so on are aspects of nature or are extranatural, still, on the whole, there is no way of determining the objective truth or falsehood of worldviews, as I mentioned above. The truth of a worldview, beyond its possible logical truth, is purely subjective. Thus all worldviews are a priori equally valid, if indeed the concept of validity is at all relevant to worldviews.

As for myself, I will reject any worldview that is at odds with science, or is logically inconsistent. But that's a consequence of my own worldview. If someone chooses to ignore science, that might have low survival value for that believer, but by what higher truth is he or she to be judged? And if someone holds inconsistent beliefs, by what higher standard is consistency a virtue?

That's not to say all worldviews are equally simple, satisfying, useful, and so on, but these are matters of individual taste and preference. In our present investigation I am not a wholly unbiased participant, and I am glad to state that my own worldview is nontranscendent. It seems to me that transcendent worldviews are intrinsically more complicated than nontranscendent ones, in the sense that the former carry the extra baggage of extranaturality. I don't see how transcendentists can deny that. But then they can make the counterclaim that transcendent worldviews are intrinsically richer and more meaningful than nontranscendent ones, and that nontranscendentists deny themselves the richer and more meaningful experience afforded by transcendent worldviews. Richer? Perhaps. More meaningful? That's a matter of individual opinion.

Nevertheless, I don't give in, and I respond with the countercounterclaim that I'm far from exhausting the wealth of experience afforded by nontranscendentism, and that I somehow manage to find enough meaning in it to keep happy. Indeed, by the principle of economy (or, parsimony), which is part of my own worldview, I much prefer making the most of the least, taking minimal conceptual raw material as far as it will go before claiming need for more. And there is still so much to be found in nature that I feel I'm wandering in a veritable wonderland. What else could anyone want? Who could ask for anything more? Well, transcendentists do, obviously.

ℛ

Metaphysics, in the sense of the philosophical framework in which science operates, is concerned with what lies around, below, above, before, and beyond science. Within the domain of science, scientists should stick to scientific considerations as far as possible. Metaphysics should consider only subjects lying outside this domain. Worldviews are conceptual frameworks and lie within the domain of metaphysics. In contrast to science, metaphysics, including worldviews, lacks criteria for objective truth, and personal taste reigns supreme. Transcendent worldviews involve the existence of a reality beyond nature. Nontranscendent worldviews make do with nature as all there is. Life, mind, consciousness, feeling, and so on fit into nontranscendentism as properties that can develop naturally in sufficiently complex systems. We compared transcendentism and nontranscendentism.

Bibliography

Some suggested books on metaphysics are

E. Conee and T. Sider, *Riddles of Existence: A Guided Tour of Metaphysics* (Oxford University Press, New York, 2007).

A. Gianelli, K. Kennedy, and G. Statile, eds., *The Journey of Metaphysics*, 2nd ed. (Pearson, Boston, 2006).

M. Lange, *An Introduction to the Philosophy of Physics: Locality, Fields, Energy, and Mass.* (Wiley-Blackwell, Hoboken, N.J., 2002).

P. van Inwagen, *Metaphysics,* 3rd ed. (Westview, Boulder, Colo., 2008).

Concerning science and religion, see

A. H. Cromer, *Connected Knowledge: Science, Philosophy, and Education* (Oxford University Press, Oxford, 1997).

P. C. W. Davies, *God and the New Physics* (Simon and Schuster, New York, 1983), a book that discusses also life, mind, consciousness, etc.

H. Fritzsch, *The Creation of Matter: The Universe from Beginning to End* (Basic Books, New York, 1984).

J. C. Polkinghorne, *Belief in God in an Age of Science* (Yale University Press, New Haven, Conn., 1998).

———, *One World: The Interaction of Science and Theology* (Princeton University Press, Princeton, N.J., 1986).

M. Shermer, *Why People Believe Weird Things: Pseudoscience, Superstition, and Other Confusions of Our Time* (Freeman, San Francisco, 1997).

J. S. Trefil, *The Moment of Creation: Big Bang Physics from Before the First Millisecond to the Present Universe* (Charles Scribner's Sons, New York, 1983).

Among these works, Davies shows that science, specifically physics, can now seriously tackle what were formerly metaphysical questions. He expresses the provocative opinion that science offers a surer path to God than does religion. Polkinghorne, a physicist turned ordained minister, discusses the differences and similarities of science and religion. I can't resist quoting Polkinghorne's beautifully understated view of Davies in *One World*: "Paul Davies, who cannot readily be suspected of being unduly influenced by Christian theology, ..."

The following two books deal with the development of life from inanimate matter through random processes obeying the laws of nature. Their main import is that the combination of chance and the laws of nature leads to inevitability; no extranatural factors are needed. Related metaphysical issues are also discussed.

M. Eigen and R. Winkler, *Laws of the Game: How the Principles of Nature Govern Chance* (Knopf, New York, 1981, and Princeton University Press, Princeton, N.J., 1993).

J. Monod, *Chance and Necessity: An Essay on the Natural Philosophy of Modern Biology* (Knopf, New York, 1971).

For a discussion of, among other things, self-organizing systems, including living systems, see

J. D. Barrow, *The World within the World* (Oxford University Press, Oxford, 1988).

I. Prigogine and I. Stengers, *Order out of Chaos: Man's New Dialogue with Nature* (Bantam Books, New York, 1984).

The latter includes a lot of metaphysical considerations.

CHAPTER FIVE

Our Unique Universe

The Lawless Universe

IN CHAPTERS 2 AND 3 we saw that in doing science we operate as follows. We investigate whatever reproducible aspects of nature interest us and look for order. From order we formulate laws, which allow prediction. Then we try to develop theories that explain those laws. The raw material of science is the reproducible phenomena of nature; they are the grist for science's mill. If some aspect of nature, however interesting it might be, exhibits no reproducibility, then it will lie outside the domain of concern of science, and the concepts of order, law, predictability, and theory will be completely irrelevant to it.

Some examples of more or less known kinds of irreproducible phenomenon, if they really exist, are parapsychology, "anomalous events," "transient phenomena," and miracles. Parapsychological phenomena include extrasensory perception (ESP), telepathy, telekinesis, and clairvoyance. Many controlled experiments have been performed to investigate those effects. To the best of my knowledge, the results of the experiments so far undertaken do not absolutely exclude such effects, but if they exist, they are weak and have yet to exhibit any reproducibility. In addition, the effects tend to weaken and vanish as the experimental controls are made increasingly rigorous.

Miracles, if one believes in them, are irreproducible by their very nature. Thus, nothing more need be said about them. Those phenomena called "anomalous events" or "transient phenomena" are peculiar, unexpected, unexplained, and definitely irreproducible occurrences. They are like miracles, but are unexpected and seem to rest upon no purpose. For example, one goes to bed at night as usual and wakes up the next morning in a strange bed in a strange city. Or one's TV, all of a sudden and for a few seconds, picks up a broadcast from another country halfway around the world. Who hasn't heard or read of something along these lines, especially with the help of the sensation tabloids at supermarket checkouts? But again, such phenomena, if they aren't merely hoaxes, are irreproducible.

So even if such interesting but irreproducible phenomena as parapsychology, "anomalous events," and miracles do exist, they will be immune to our *scientific* understanding (although they might be amenable to nonscientific modes of comprehension and understanding). Order, law, predictability, and theory, in the sense of science that we discussed in chapters 2 through 4, will be irrelevant to them, and *in the sense of science* they will justifiably be described as orderless, lawless, unpredictable, and unexplainable.

Although science cannot comprehend irreproducible phenomena, this fact does not preclude their occurrence. One might or might not believe that such phenomena occur in the material universe with which we can, or can conceivably, interact, but their incompatibility with the framework of science should not serve as the rationale for anyone's disbelief. Science, as our *attempt* to understand objectively the reproducible and predictable aspects of nature, is not claimed to be the final arbiter of the occurrence of anything.

Stated in other words: Who is to say that all of nature must be reproducible? Not science! Science aims strictly at those aspects of nature that have been found to be reproducible, and ignores nature's irreproducible aspects. Thus science has nothing whatsoever to say about whether nature possesses any irreproducible aspects or not. For example, telepathy might be a real effect. The inability of science to comprehend telepathy, owing to its irreproducibility, is in itself no reason not to be open-minded about its possible existence. (Science does, however, offer other reasons for attributing a low likelihood to telepathy.)

Where are reproducible phenomena found in nature? Or, in other words, where in nature can we search for order, law, predictability, and explanation? We know from experience that sufficiently small systems exhibit reproducible behavior, where sufficiently small is more or less human size and smaller. (Actually, to be more precise, such systems are found to be reproducible when they are isolated from their surroundings in order to prevent the effect of uncontrollable external influences.) These systems (discussed in more detail in *Quasi-Isolated System and Surroundings*, chapter 6) are the systems we can manipulate and experiment with. Reproducibility is discovered by repeating our experiments (*Reproducibility*, chapter 2). And to be able to experiment and repeat, we have to be able to manipulate. As an ordinary example, we find reproducibility in our own bodies. We can reproducibly tie our shoes, button our buttons, and grasp the objects we intend to. But for more precise investigation, we perform laboratory experiments, in which we find and study reproducible phenomena.

Systems much larger than human size cannot be manipulated, but such systems might be considered reproducible by declaration. For example, geological phenomena, such as earthquakes, mountains, glaciers, and volcanoes, are not reproducible at will in the laboratory. They are, however, treated as reproducible in the sense that nature supplies us with a sufficient variety and frequency of them that we can consider nature to be manipulating for us and presenting us with performed experiments at size scales beyond our manipulative capabilities. Such reasoning extends to even larger systems: planets, planets' moon systems, stars, planetary systems, galaxies (groupings of millions of stars), galaxy clusters (groupings of galaxies), and galaxy superclusters (clusters of galaxy clusters). But the grounds for considering them reproducible weaken toward the end of the list, since the larger the system, the less the variety and the lower the frequency that nature presents us for observation.

As we progress from small, readily reproducible systems to systems of increasing size, actual reproducibility begins to become invalid, to be replaced by declared reproducibility, which in turn loses its justification. When we push matters to their extreme and consider the whole universe, we have clearly and irretrievably lost the last vestige of reproducibility; the universe as a whole is a *unique phenomenon* and as such is intrinsically irreproducible.

Let's look into this point in more detail. Certainly we can't set up universes at will. But perhaps the universe we're familiar with can be *declared* reproducible, since theoretical physicists have put forth proposals involving the existence (in some sense) of unimaginable numbers of other universes. Well, it cannot thus be made reproducible. As we'll see in the following samples, in all these proposed universe ensembles there is no interaction, not even a conceivable possibility of interaction, among universes. So we in our universe cannot, even conceivably, interact with any of the others. In all these cases, therefore, the proposed other universes are no part of nature and do not interest us in our present investigation. (Recall that nature is the material universe with which *we can, or can conceivably, interact*.) So as far as science is concerned, our universe remains unique. The proposed ensembles are metaphysical, not natural ones. Here are three samples.

First: There is a most successful theory, called *quantum theory*, that is concerned with the fundamental behavior of all systems in principle, but is usually and most usefully applied to molecular, atomic, and subatomic systems. The theory is formulated in terms of possible happenings and their probabilities of actually occurring (rather than, as for Newton's theory of Kepler's laws, in strictly deterministic terms of what *will* occur). The theory itself is formal and mathematical. Although it is universally accepted among scientists, since it always gives excellent results, its conceptual interpretation is still the subject of much controversy.

One interpretation of quantum theory, called the many-worlds interpretation, proposes that at every instant the universe "branches" into realizations of all the quantum possibilities of that instant, which continue to "coexist side-by-side," each branch universe branching further at the next instant, and so on to unimaginable profusion.

Thus, by this proposal, there is no fundamental distinction between what actually happens and what might have happened. Every possibility is realized in its appropriate branch universe, while we happen to find ourselves in only one of the branch universes and perceive reality accordingly. Parallel observers in other branch universes observe matters differently from us. They observe realities that for us could have developed but did not, while we experience what is for them an unrealized possibility.

These "side-by-side, coexisting" universes are supposed, by the many-worlds interpretation, not to interact with each other. So we in our branch universe cannot, even conceivably, interact with any of the other branch universes. That places all the other branch universes outside nature for us. Thus, the many-worlds picture is not a description of nature, but is rather a metaphysical idea. The proposed "side-by-side coexistence" is ponderable only in a purely metaphysical sense. Science can offer no answer to the question "Where did I win?" asked by the loser of a coin toss.

Second: The proposed "cosmic oscillation" scheme of the evolution of the universe has the universe come into being in a "big bang" (a cosmic explosion), expand to maximal extension, contract down again, and end its life in a "big crunch" (a cosmic implosion). That is followed by the big-bang birth of another universe, its expansion, contraction, and final demise in a big crunch. And so on and on, with no beginning and no end. No influence or information is supposed, according to this scheme, to carry over from one universe to the next. The noninteraction among universes makes this ensemble, too, a metaphysical one, since we in our universe can't interact with any of the others. As far as science is concerned, this "sequence" of universes is no temporal sequence, because ordinary time has meaning only in relation to our own universe. It can be thought of as a sequence only in some metatime, a purely metaphysical construct. Questions such as "When is the universe in which I have blue eyes rather than brown?" have no meaning within the framework of science.

Third: The big-bang scheme for the biography of the universe has the universe come into being some 14 thousand million years ago in the form of a cosmic explosion, a primeval fireball of extreme density, pressure, and temperature that has been expanding and cooling ever since. This scheme has a version in which, during an era of unimaginably rapid expansion—called the inflationary era—starting soon after the big bang, the expanding universe broke up into island universes, one of which is ours. Those universes have been carrying on the cosmic inflation by flying away from each other at such stupendous speeds that no interaction among them has been possible, since no influence emanating from any island universe could propagate fast enough to overtake any other. They are now supposed to be so far apart and dispersing so rapidly that still no interaction among them is possible or ever will be.

One might feel that this third universe ensemble has more of a natural flavor to it than do the previous two. At least I do. I can imagine something like many universes flying apart from one another, though I can't at all imagine the other two schemes. But even in this scheme the noninteraction puts the other universes once again outside the framework of science and renders the ensemble metaphysical, since we in our island universe can't interact with (or even observe) any of the others. There is no scientific answer even to the question of how far and in what direction our nearest-neighbor island universe is supposed to be.

So, being a unique phenomenon, the universe is intrinsically irreproducible. Thus the universe *as a whole* lies outside the framework of science. Order, law, predictability, and explanation are irrelevant and inapplicable to it, and we justifiably call it orderless, lawless, unpredictable, and unexplainable. Or, using livelier language, as far as science is concerned the universe does as it damn well pleases.

The universe, then, is the limit of our possible scientific understanding of the material world, while it itself, as a whole, can have no explanation in science. As for *scientific* understanding of the working of the universe *as a whole,* we will never be able to state anything more enlightening than it is because it is.

Cosmology

Cosmology is the study of the working of the cosmos, the universe as a whole. But since the universe as a whole lies outside the framework of science, it would seem that cosmology is really a branch of metaphysics, rather than, as it is often considered to be, a branch of science. The unpredictable, orderless, irreproducible, unique universe certainly exhibits aspects and phenomena that possess reproducibility, order, and predictability. And we certainly can and do explain aspects of and phenomena within the universe on the basis of other aspects and phenomena of it. As long as cosmology deals with the connections and interrelations among those aspects and phenomena, it is a branch of science. But in its holistic mode, when it attempts to comprehend the universe as a whole, as it is intended to do, cosmology can only be a branch of metaphysics.

Let's see how cosmology operates. Cosmology attempts to describe the working of the universe at present, in the past, and in the future on the basis of all data available to us here and now. These data are the laws of nature known to us, the present material composition and state of the universe in our cosmic vicinity (say, the region of the solar system), and information obtained by telescopic means (giving data only about the past, due to the finite speed of light, of radio waves, and of other carriers of information). But to construct a scheme of the working of the universe, we must make assumptions that are completely unverifiable, owing to the intrinsic irreproducibility of the universe as a whole.

For example, we might, and invariably do, assume that the laws of nature known to us now were and will continue to be valid also as far into the past and future as our scheme requires. This assumption makes good, conservative science practice, but it is unverifiable and is thus a metaphysical assumption. Or, we might, and also invariably do, additionally assume that the laws of nature known to us here in our cosmic vicinity are valid also everywhere else in the universe. Again, this assumption, although good, conservative science practice, is unverifiable, so it too is a metaphysical assumption.

Some claim they can confirm the uniformity of the laws of nature in the past and over space, on the basis of telescopic observations, especially of stellar spectra. They may be right. But the future remains an unknown, and their conclusions, like all experimental conclusions, depend on how their data are interpreted. Their interpretation depends on cosmological assumptions, which are components of cosmological schemes. (More on that later.) The construction of a scheme by assuming it is circular reasoning. It is likely that a self-consistent scheme involving uniform laws of nature is achievable. But unverifiable assumptions will be involved nevertheless.

Whatever is assumed in these and in many other regards, the most we logically must demand of a scheme of the working of the universe is that it be consistent with all the data and that it be self-consistent, that is, contain no internal contradictions. Because these requirements might not be sufficient to determine a unique scheme, it might happen that more than one self-consistent scheme of the working of the universe fits all the data. There is no guarantee for the uniqueness of a consistent scheme, nor does science offer any criterion of preference among consistent schemes. Thus any criterion of preference, such as simplicity, beauty, or conservatism, is necessarily metaphysical. (That is true, just as well, of criteria of preference for theories, as we discussed in chapters 3 and 4.)

Cosmological schemes should not be confused with theories, however. Cosmological schemes are concerned with an intrinsically irreproducible phenomenon, our unique universe, whereas theories are concerned with laws expressing order among reproducible phenomena. Cosmological schemes are attempts to *describe* the working of the universe, not to *explain* it scientifically, since it is intrinsically scientifically unexplainable. And they are basically metaphysics, since, due to the intrinsic irreproducibility of the universe, they must contain unverifiable assumptions. So if cosmology is basically metaphysics, what is *cosmogony*, which deals with considerations concerning the *origin* of the universe, if not almost pure metaphysics all the more?!

That is *not* to say that cosmology is a waste of time! Cosmological schemes, however metaphysical they are and must be, are of immense value for our attempt to understand the universe. They offer insight into the connections and interrelations among the aspects and phenomena that make up the universe as a whole. They guide us in directing our investigations and give us a framework for organizing our data. They supply terms of reference for formulating laws and devising theories. And they might even have predictive power, thus exposing themselves to falsification.

To see that in action, consider, for instance, cosmological schemes of the type currently under intensive investigation, the inflationary big-bang schemes (briefly outlined in the previous section). These schemes of the universe are wide-ranging, encompassing aspects from the largest scale to the smallest, from the spatial distribution of galaxy superclusters to the properties and interactions of elementary particles. The schemes describe how these aspects mesh and affect each other, combining to make up the whole. The schemes include effects and phenomena that have not yet been observed, such as the future evolution of the universe, extra-large-scale astronomical structure (involving galaxy clusters and superclusters), and as-yet undiscovered kinds of elementary particles and properties of the elementary-particle interactions. Thus the schemes guide research in these areas, provide a framework for interpreting and correlating results, and even offer predictions.

Cosmological schemes describe the working of the universe, and so retrodict (i.e., tell about the past) the previous evolution and even the origin of the universe, have something to say about its present state, and predict its future evolution. As such, they might appear to be laws of behavior for universes. Yet cosmological schemes are not and *cannot* be laws of behavior for universes, or any other kinds of law for that matter. They were *not* obtained by investigating the behavior of reproducible systems. We can't run experiments in universe evolution, and nature does *not* supply us with more than the single universe we have. (At least that's how things seem at present, and we don't see any prospect that the situation will be any different in the future.) Thus there is no reason at all to think that cosmological schemes describe what *really* happened or will happen.

As for the apparent retrodictive power of cosmological schemes, recall that such schemes are based on the available data. These data in themselves do not tell us the past behavior of the universe. Not even do the data obtained by telescopy tell us the universe's past behavior, even though these data refer to the past, owing to the finite speed of light and radio waves. The data must be interpreted to be meaningful. But it is just such cosmological schemes that serve as tools for interpreting data. And we have no independent way of finding out what really happened. There is no Rip van Winkle who, after sleeping a few billion years, can wake up and tell us what was going on in the universe before he fell asleep. We can only interpret the data we find here and now, and we need cosmological schemes for that.

As an example, consider the astronomical observation known as the red-shift effect, which is that the stars in distant galaxies appear redder than those in closer ones. What does this effect tell us about the past, when the light left the stars to travel unimaginably vast distances at the speed of 300,000 kilometers (186,000 miles) per second until it reached our telescopes? Even the idea that the light we receive today is the same light that left the stars so long ago is an assumption based on some cosmological scheme telling us what happens to light on its way from stars to us. It could be, for instance, that the original starlight was absorbed by dilute interstellar matter lying in its path, which then emitted new light, redder than the light it absorbed. In that case, the red shift would be telling us something about the properties of interstellar matter at the time the light encountered it.

But if we assume that the light reaching our telescopes is actually the same light that left the stars, then the red shift might be telling us that stars were really redder in the past, since the light reaching us now left the farther stars before it left the nearer ones. Or the red shift might be telling us that stars were always the same color, more or less, but their galaxies are and were moving away from us, the more distant galaxies receding faster than the closer ones. The reddening of the light would then be the result of the "stretching of space," according to Einstein's general theory of relativity. In less technical discussions, the reddening is often attributed instead to the well-known Doppler effect (after Christian Johann Doppler, 1803–1853), by which light appears redder or bluer, or a tone sounds lower or higher in pitch, when its source recedes or approaches, respectively. (Recall the changes in pitch of an ambulance or police siren or train whistle as the vehicle approaches and then speeds by.)

This interpretation of the red-shift data is the one most commonly accepted. It is based on the cosmological assumption that all the galaxies in the universe are receding from each other, which is consistent with the big-bang cosmological scheme, by which the universe has been expanding since it came into being. The main point for us, however, is not this cosmological scheme or that, but rather, if the data are to tell us something significant about the past, they must be interpreted by means of *some* cosmological scheme or other.

So the apparent retrodictive power of cosmological schemes turns out to be illusory. Cosmological schemes paint dynamic pictures of the universe's past, perhaps even its origin, on the basis of everything known here and now, but we have no independent source for confirmation. To put the matter bluntly, as regards telling us the origin of the universe and its past evolution, cosmological schemes are actually no better than fairy tales.

Cosmological schemes also predict the future evolution of the universe. We have seen that these schemes are not laws, hence there is absolutely no compelling reason to believe that the future they ascribe to the universe is what will really occur. Yet it would seem that simply by waiting patiently it should be possible for us to compare their predictions with reality. But patience won't help. The time scales involved in the predicted changes are so enormous that the predictive power of cosmological schemes, although valid in principle, is utterly nonexistent in practice. To emphasize the point, the Sun can be expected to explode and broil the solar system long before any predicted evolution should become evident. So, as regards telling us the future evolution of the universe, cosmological schemes fare scarcely better than when they pretend to tell us its past.

Earlier in this section I stated that cosmological schemes, among their other useful properties, might have predictive power, thereby exposing themselves to falsification. In the meantime we have seen that their predictive power involving the past or the future is merely an illusion. The nonillusory remainder is their predictive power regarding the *present* state of the universe. Here lies their opportunity to shine. Recall that such schemes are based on the available data. Cosmologists might then predict something about the universe as it is right now, something that can be tested by performing experiments now, obtaining new data, and comparing these data with what the schemes say they should be.

For example, assuming it is not known how galaxy superclusters are distributed throughout space (this is currently being investigated), a cosmological scheme might predict that the spatial distribution of superclusters should be uniform. Astronomers can then perform their observations and analyses and come up with data about supercluster distribution, determining whether they appear to be distributed uniformly or not. (The latter seems to be the case.) Thus the cosmological scheme can be tested. Or, a cosmological scheme might predict the existence of a previously unknown kind of elementary particle or property of elementary-particle interaction. Then the particle investigators can get to work and produce data about the existence or nonexistence of the predicted kind of particle or interaction property. And again the scheme can be tested.

So the real predictive power of cosmological schemes, with its concomitant falsifiability, is confined solely to the present. Whatever such schemes have to say about the past and future of the universe should be recognized as illusory from the point of view of science, and relegated to the science fiction shelf.

The sweep and range of cosmological schemes can be so exhilarating that some theoretical physicists have expressed their conviction (or at least their feeling) that eventually, with sufficient ingenuity and diligence, it should be possible to develop a Theory of Everything (acronymed to TOE). As we have learned in this chapter, a TOE cannot be a real theory; it would be a cosmological scheme. It would not be a scientific explanation of the universe, but rather a description of it. Yet its descriptions of the origin, past, and future of the universe would be unfounded from the scientific point of view. The TOE's predictive power would be confined solely to the present state of the universe.

Still, were a TOE ever developed, it would certainly be a wonderful accomplishment of immense value, supplying a unifying framework and insight for our attempts to fathom the connections and interrelations among the aspects and phenomena that are components of the universe as a whole. Nevertheless, and to add some perspective, there are also those theoretical physicists who believe that a TOE is unattainable, either in principle or in practice.

ℛ

The universe *as a whole* is a unique phenomenon as far as science is concerned. It is therefore intrinsically irreproducible, and thus lies outside the framework of science. Thus order, law, predictability, and theory are irrelevant to the universe as a whole; it is *orderless, lawless, unpredictable,* and *unexplainable.* Hence cosmology is a branch of metaphysics, and cosmological schemes, such as the inflationary big-bang schemes, are not theories; they are attempts to *describe,* not to *explain* scientifically, the working of the universe. Nevertheless, cosmological schemes are of immense value for science, offering insight, guidance, frameworks, and predictions. Such schemes' apparent predictive power concerning the origin and past and future evolution of the universe is illusory, and their real predictive power is confined to the present.

Bibliography

Here is a list of books whose sole or main subject is cosmology. I suggest that a reader of one of these works try to discern whether the author is assuming that the universe as a whole is subject to law (which would not be science) or not. Also, try to discern the unverifiable assumptions entering whatever cosmological scheme is being presented. Is the scheme being presented as a scientific theory (which it cannot be)? Is the scheme being presented as The Way Things Were, or only as what seems to the author to be the most reasonable interpretation of the available data?

J. D. Barrow and J. Silk, *The Left Hand of Creation: The Origin and Evolution of the Expanding Universe* (Heinemann, London, 1983).

P. C. W. Davies, *The Edge of Infinity* (Simon and Schuster, New York, 1981).

————, *The Runaway Universe* (Harper and Row, New York, 1978).

T. Ferris, *The Whole Shebang: A State-of-the-Universe(s) Report* (Simon and Schuster, New York, 1997).

H. Fritzsch, *The Creation of Matter: The Universe from Beginning to End* (Basic Books, New York, 1984).

A. Guth, *The Inflationary Universe: The Quest for a New Theory of Cosmic Origins* (Addison-Wesley, Reading, Mass., 1997).

R. Morris, *The Fate of the Universe* (Playboy Press, New York, 1982).

H. R. Pagels, *Perfect Symmetry: The Search for the Beginning of Time* (Simon and Schuster, New York, 1985, and Bantam, Toronto, 1986).

J. Silk, *The Big Bang,* revised and updated ed. (Freeman, San Francisco, 1989).

J. S. Trefil, *The Moment of Creation: Big Bang Physics from Before the First Millisecond to the Present Universe* (Charles Scribner's Sons, New York, 1983).

S. Weinberg, *The First Three Minutes: A Modern View of the Origin of the Universe* (Basic Books, New York, 1977).

Now a number of books that are concerned with more than just cosmology, but that contain material on cosmology. My suggestions hold for these too.

R. K. Adair, *The Great Design: Particles, Fields, and Creation* (Oxford University Press, Oxford, 1987).

J. D. Barrow, *The World within the World* (Oxford University Press, Oxford, 1988).

F. Close, *The Cosmic Onion: Quarks and the Nature of the Universe* (American Institute of Physics, New York, 1983).

P. C. W. Davies, *God and the New Physics* (Simon and Schuster, New York, 1983).

————, *Space and Time in the Modern Universe* (Cambridge University Press, Cambridge, 1977).

B. Greene, *The Fabric of the Cosmos: Space, Time, and the Texture of Reality* (Vintage, New York, 2004).

————, *The Elegant Universe: Superstrings, Hidden Dimensions, and the Quest for the Ultimate Theory* (Vintage, New York, 2003).

J. Gribbin, *The Search for Superstrings, Symmetry, and the Theory of Everything* (Back Bay, Newport Beach, Calif., 2000).

S. W. Hawking, *A Brief History of Time: From the Big Bang to Black Holes* (Bantam, London, 1988).

R. Morris, *The End of the World* (Anchor Press, Garden City, N.Y., 1980).

D. Park, *The How and the Why* (Princeton University Press, Princeton, N.J., 1988).

J. S. Trefil, *Reading the Mind of God: In Search of the Principle of Universality* (Charles Scribner's Sons, New York, 1989).

S. Weinberg, *Dreams of a Final Theory* (Pantheon Books, New York, 1993).

The following book surveys the history of cosmogonical ideas, from ancient "creation myths" to modern thinking on the origin of the universe.

M. Gleiser, *The Dancing Universe: From Creation Myths to the Big Bang* (Dutton (Penguin), New York, 1997).

For quantum theory (science) and its many-worlds interpretation (metaphysics), see

P. C. W. Davies, *Other Worlds: A Portrait of Nature in Rebellion; Space, Superspace and the Quantum Universe* (Simon and Schuster, New York, 1980).

P. C. W. Davies and J. R. Brown, eds., *The Ghost in the Atom: A Discussion of the Mysteries of Quantum Physics* (Cambridge University Press, Cambridge, 1986).

R. P. Feynman, *The Character of Physical Law* (MIT Press, Cambridge, Mass., 1965).

T. Hey and P. Walters, *The Quantum Universe* (Cambridge University Press, Cambridge, 1987).

H. R. Pagels, *The Cosmic Code: Quantum Physics as the Language of Nature* (Simon and Schuster, New York, 1982).

J. C. Polkinghorne, *The Quantum World* (Princeton University Press, Princeton, N.J., 1984).

A. Rae, *Quantum Physics: Illusion or Reality* (Cambridge University Press, Cambridge, 1986).

F. Rohrlich, *From Paradox to Reality: Our New Concepts of the Physical World* (Cambridge University Press, Cambridge, 1987).

Also, from the lists above, Adair, Barrow, Davies (1983), Fritzsch, Hawking, and Park.

For other universes (metaphysics), see

J. Gribbin and M. Rees, *The Stuff of the Universe: Dark Matter, Mankind and the Coincidences of Cosmology* (Heinemann, London, 1989).

L. Smolin, *The Life of the Cosmos* (Oxford University Press, New York, 1997).

Also, from the lists above, Barrow and Silk, Davies (1978), Ferris, and Morris (1980 and 1982).

For putative TOEs, Theories of Everything (which would not be scientific theories), see

P. W. Atkins, *The Creation* (Freeman, San Francisco, 1981).

P. C. W. Davies and J.R. Brown, eds., *Superstrings: A Theory of Everything?* (Cambridge University Press, Cambridge, 1988).

Also, from the lists above, Adair, Gribbin and Rees, Hawking, and Weinberg (1993).

Some books on irreproducible and unpredictable phenomena are

H. Broch, *Exposed! Ouija, Firewalking, and Other Gibberish* (Johns Hopkins University Press, Baltimore, 2009).

G. Charpak and H. Broch, *Debunked! ESP, Telekinesis, and Other Pseudoscience* (Johns Hopkins University Press, Baltimore, 2004).

M. Shermer, *Why People Believe Weird Things: Pseudoscience, Superstition, and Other Confusions of Our Time* (Freeman, San Francisco, 1997).

CHAPTER SIX

Nature's Laws

IN *THE LAWLESS UNIVERSE,* chapter 5, we saw that the universe *as a whole,* being a unique phenomenon, is irreproducible, orderless, lawless, unpredictable, and unexplainable by science. *Within* the universe, however, we find aspects and phenomena that *are* reproducible, orderly, lawful, predictable, and explainable by science. Since aspects of the universe and phenomena within it are parts of the whole universe, the behavior of these parts is a constituent of the behavior of the whole. Their behavior can even be thought of as being engendered by the behavior of the whole. Thus their behavior might be expected to be as irreproducible, orderless, lawless, unpredictable, and unexplainable as the behavior of the whole. That notion derives from our common experience that parts of a totally unruly situation are generally every bit as unruly as the whole, if not more so.

The question thus arises: How is it that within the lawless universe we find laws of nature? The resolution of this apparent contradiction is the principal subject of the present chapter.

Realism and Idealism

On the way to our destination, to lay a foundation and broaden understanding, we'll pass through and linger at a few stations. Our first stop is a consideration of the question: Where do the laws of nature reside?

Nature, at least in certain of its aspects (such as planetary motion, as we saw in chapters 2 and 3), "behaves lawfully," "obeys laws." We discover those laws. Do the laws then reside in our recognition of them? Would the laws not exist if we didn't discover them?

Did Newton's law of gravitation (*An Archetypal Example*, chapter 3) not exist before Newton? Still, we would expect nature to behave in the same way whether we discovered its laws or not. Yet could that behavior be "lawful" even if we didn't recognize it as such? If it could be, then the laws of nature would reside in nature independently of ourselves. But if there is no way that nature could be considered to obey unrecognized laws, then in some sense the laws would have to reside in ourselves.

Consider again our experiment of a sphere rolling down an inclined plane (*Predictability*, chapter 2). Without a doubt, the distance the sphere rolled from rest was proportional to the square of the time interval even before we discovered that law. There's no reason to suspect nature of behaving differently when we observe it than when we don't. (In reality, nature *does* behave differently when we observe it and when we don't, as described by quantum theory. But that distinction is significant only on the submicroscopic scale, and is undetectable on the scale of spheres rolling down planes.) Nature was just doing its thing all along. Was its behavior lawful before we determined it was?

So where do the laws of nature reside? Are they wholly "out there," in the objective real world external to the observer? Or are they completely "in here," subjective constructs of the observer's mind? Or some of both? The question is pure metaphysics. The former approach is the metaphysical position called *realism*, which is held by virtually all scientists, although almost always unconsciously and without considering other possibilities. This worldview has it that the laws of nature are "really" there, independent of observers, that they are objective and would have reality even if we weren't around to discover them, and that they would be discovered just the same by any kind of sufficient intelligence.

The process of discovery, according to the realist position, is "merely" the recognition of the validity of the laws. Thus Newton's laws (*An Archetypal Example*, chapter 3), for example, existed and were valid even before Newton discovered, or recognized, them. Realism is part of the spirit of most of science. Although realism does not demand conservatism, it does encourage it. In regard to laws of nature, conservatism takes the form: As long as there is no compelling reason to the contrary, assume that the laws of nature we find here and now are, were, and will be valid everywhere and forever.

Regardless of one's political views, conservatism is generally considered good science practice. (That is a metaphysical position.) In broad terms, the conservative platform in science is this. Hold on to what you have, stick to the tried and well-confirmed, for as long as is reasonably possible; make changes only when the need for change becomes overwhelming; and then make only the minimal changes needed to achieve the desired end. The geocentric model of the universe might serve as an example of that. It certainly appeared valid. Everything in the sky seemed to be revolving around Earth. There was no compelling reason to give up the picture in favor of Nicolaus Copernicus's (1473–1543) heliocentric model, with Earth and the other planets revolving around the Sun. Until Galileo Galilei (1564–1642) used his improved telescope and observed the phases of Venus (like the phases of the Moon) and moons of Jupiter. The evidence that Earth was no different than the other planets and was not at the center of things eventually became overwhelming, and the heliocentric model came to be accepted.

Conservatism demands, among other things, that, given the choice, one prefer theories that assume that the laws of nature we find here and now are, were, and will be valid everywhere and forever, in preference to theories that assume otherwise. If one held that the laws of nature were not wholly "out there," were not well anchored in the objective concrete of external reality, one would more likely be tempted (as long as one is theorizing anyway) to assume variable (in time and/or in space) laws of nature than if one held a realist position. Or so it seems to me. That's how realism encourages conservatism.

The metaphysical position that the laws of nature are wholly "in here," in the mind of the observer, is called *idealism*. This worldview is rare among scientists. Idealism says the laws of nature we "discover" are not intrinsic to the "raw material" of the external world, but are mental constructs, artifacts of the way our minds interpret and organize our sensory impressions, of the way we perceive the world. Thus our laws of nature are determined by the properties of our sense organs and by our mental makeup.

Intelligent beings with a mental makeup drastically different from ours might interpret and organize their perceptions so as to discover laws of nature so wildly different from ours, yet valid for the world as they perceive it, that their laws would be irrelevant to the world as we perceive it. We might wonder if they and we would be able to communicate at all, or even to recognize intelligence in each other.

Between the poles of realism and idealism lies a range of possibilities for other metaphysical positions concerning the seat of the laws of nature. One such hybrid worldview, which I find especially attractive, is that there indeed is *order* "out there" in the world external to ourselves, and that intelligent beings perceive it via their senses and formulate laws from those perceptions in accord with their mental makeups. Order would thus be objective, whereas laws of nature would be species-subjective, perhaps even individual-subjective.

According to this position, intelligent beings evolve and develop within, and by, the objective order. So their senses and mental makeups must have survival value and cannot bring about behavior that is too "unnatural," such as acting according to "order" that has no objective support, or ignoring orderly phenomena that affect the beings strongly.

As an example of acting according to nonorder, imagine a being on Earth who somehow forms the "law" that whenever a light source is present in the sky (Sun, Moon, or stars) a tornado will strike imminently, and it's essential to bury one's head in the sand to avert danger. Such a being would have a low chance of surviving, mainly since there would remain little time for it to look for food, and secondarily since it wouldn't be well protected should a tornado actually happen to strike.

As regards ignoring important objective order, consider an imaginary being who searches for food at random times during the day and night, ignoring the fact that food is available only during the day. Such a being would both waste energy in search for nonexistent food at night and not fully exploit the food potential of the day, thus suffering a disadvantage compared to a being who acted in better accord with the objective order.

Human examples of these sorts of antisurvival behaviors are astrology (acting according to nonorder) and rejection of the insights of medical science (ignoring objective order). In the former case, a fanatically astrology-believing general would choose to give up a tactically favorable, but horoscopically unfavorable, date for attack, and instead prefer to move his or her forces on a horoscopically favorable, but tactically unfavorable, date. The result could be catastrophic (for that side in the conflict). Fortunately for astrology-inclined people, most of them act according to their horoscope only in matters where it makes little difference either way.

Rejecting the insights of medical science is often a serious matter. By ignoring the well-known dangers of smoking, smokers reduce both the quality of their lives and their life expectancies. Eaters of poor diets do the same, when they ignore what is known about the effects of diet on health. Whatever the advantages of alternative methods of healing might be, rejection of conventional medicine, as espoused by certain religious groups, can be (and on occasion is) a source of suffering and death.

Although, as we just saw, the objective order "out there" must impose certain features on the laws of nature fashioned by any viable species or individual, the hybrid position we are considering recognizes that differing senses and mental makeups can develop. That could happen where beings evolve in diverse environments, consequently forming differing worldviews and laws of nature (within the restrictions imposed by the objective order). Compare, for example, what we can reasonably infer to be the laws of nature of dolphins with those of birds, or each with ours. And, letting our imagination wander, we would hardly expect an intelligent galaxy (should a galaxy have or ever develop intelligence) to view matters in quite the same manner as we do.

You might be wondering, at this point, what all the fuss is about. Am I not splitting hairs, with order sitting "out there" and laws "in here"? After all, aren't order and laws pretty much the same thing, the laws forming descriptions of order and abstractions from order? The point, however, is that *laws predict.*

Nature presents us with a set of happenings. We find order among them and formulate laws that abstract from the order and describe it. The laws predict additional happenings; they tell us what would happen in situations that have not yet occurred. Since those situations have not yet occurred, the happenings that would happen in them have not yet happened. So the laws contain more information than nature has provided us. Where does this information come from? It comes from our minds. We inject it during the inspired passage from perceived order to laws. When we extrapolate from the actual set of situations and happenings to all possible situations and consequent happenings, we inject something of ourselves. Thus, laws are made of data from the real world *as well as of stuff from our minds.*

That's why I don't view the laws of nature as residing "out there." What *is* out there are actual happenings. The order we find belongs to the *actual* happenings; no hypothetical happenings are involved. So I place the order "out there." Laws, too, deal with *actual* happenings, but, importantly, they deal with *potential* happenings as well. Potential happenings aren't "out there" until they happen. Thus I have laws dwelling "in here."

Let's return to the rolling-sphere experiment (*Predictability*, chapter 2). There were ten actual happenings, ten rolls of the sphere. When we plotted the roll data in a certain way, they appeared to fall on a straight line. At that stage we discovered objective order in the set of happenings. We did not inject order. It was inherent to the happenings. That's why I view order as being "out there."

But when we declared that the law $d = 0.10t^2$ not only describes the *actual* rolls, which it does, but predicts the results of any number of *potential* rolls, we went beyond what nature revealed to us. Our extrapolation injected something of ourselves. What did we inject? We could have drawn a snaky curve through the ten data points instead of a straight line. That, too, would have been objective order: the ten data points appear to fall on such and such a curve. Then we could have come up with a mathematical relation between d and t that is much more complicated than $d = 0.10t^2$ and declared it a putative law. (It would not have stood the test of additional rolls, but that's not the point here.)

What, then, are we injecting? We're choosing which objective order to extrapolate into law. In the present example, we chose what we perceive to be the simplest order, the one giving the simplest mathematical relationship relating the ten data points. We prefer simplicity, and we injected our preference into the law. (As it turns out, the law is correct. Nature seems to prefer simplicity too.) That's why I view the law as residing "in here."

Reductionism and Holism

The next stop on our way to a clarification of how it is that there are laws of nature within the intrinsically lawless universe is yet another polarity, which we can add to the realism-idealism one.

We live in nature and view it and are intrigued. Our material needs and our curiosity drive us to try to understand what is happening around us. What we observe in nature is a complex of phenomena—some apparently interrelated, others seemingly independent—including ourselves, where we're related to all of nature, as is implied by our definition of nature as the material universe with which *we can, or can conceivably, interact*. The possibility of interaction is what relates us to all of nature and, owing to the mutuality of interaction and of the consequent relation, relates all of nature to us.

It follows that all aspects and phenomena of nature are interrelated, whether they appear so or not. Whether they are interrelated independently of us or not, they are certainly interrelated through our mediation. For example, let's say that aspect A of nature is not directly related to phenomenon P. But since both A and P are components of nature, we can interact with both. Thus they are interrelated by means of us. A can interact with P by interacting with us, thereby causing us to interact with P, and vice versa. In this manner all of nature, including *Homo sapiens,* is interrelated and integrated.

Science is our attempt to understand (the reproducible and predictable aspects of) nature objectively. But how are we to grasp this wholeness, this integrality? Nature, in its completeness, appears so awesomely complex, owing to the interrelation of all its aspects and phenomena, that it might seem utterly beyond hope to understand anything about it at all. True, some obvious precincts of simplicity stand out, such as day-night periodicity, the annual cycle of the seasons, and the fact that fire consumes. And subtler simplicity can also be discerned, such as the term of pregnancy, the correlation between clouds and rain, and—still more subtle—the relation between tides and the phases of the Moon.

Yet on the whole, complexity seems to be the norm, and even simplicity, when considered in more detail, reveals wealths of complexity. But owing to nature's unity, to its integrality, any attempt to analyze nature into simpler component parts cannot avoid leaving something out of the picture.

That brings us, at length, to the metaphysical position called *holism.* According to the holist worldview, nature can be understood only in its wholeness or not at all. And that includes human beings as part of nature. So long as nature is not yet understood, there is no reason a priori to consider any aspect or phenomenon of it as being intrinsically more or less important than any other. Thus it is not meaningful to pick out some part of nature as being more "worthy" of investigation than other parts. Neither is it meaningful, according to the holist position, to investigate an aspect or phenomenon of nature as if it were isolated from the rest of nature. The result of such an effort would not reflect the normal behavior of that aspect or phenomenon, since in reality it is not isolated at all, but is interrelated with all of nature, including us.

The other pole to be examined here is the metaphysical position of *reductionism.* The reductionist position is that nature is understandable as the sum of its parts, that nature should be studied by analysis, should be "chopped up" into, or reduced to, simpler component parts that can be individually understood. A successful analysis should be followed by synthesis, whereby the understanding of the parts is used to help us attain understanding of larger parts compounded of understood parts. If necessary, that should be followed by further synthesis, further compounding the compound parts to obtain even larger parts. Then an understanding of the latter can be attained with the help of the understanding achieved to that point. And so on to the understanding of ever more comprehensive parts, until an understanding of all of nature is achieved.

To compare the two viewpoints, consider the phenomenon of life. A reductionist would say that a living organism can be understood by understanding the workings of its smallest components, and then gaining knowledge of progressively larger components by building up from each level to the next larger one. We should start by investigating atoms and how they combine to form molecules. Then we should investigate how molecules combine to form macromolecules. Then how the latter combine to form cell components, such as membranes. And finally how such components combine to form a whole cell. Then we would understand the living cell.

A holist, on the other hand, would claim that a living cell is more than merely the sum of its components, that life is a property of the cell *as a whole* and cannot be understood in terms of parts. The most that can be understood by the reductionist method, a holist would claim, is a dead cell.

Between the poles of reductionism and holism stretches a range of positions with regard to the separability of nature. Though it is debatable if and to what extent one's position along the realism-idealism range affects the way one does science (and there are arguments in both directions), it is clear that one's position within the reductionism-holism polarity should strongly influence the way one does science, and even whether one does science at all. Each of the poles, holism and reductionism, has a valid point to make. Nature is certainly interrelated and integrated, at least in principle, and we should not lose sight of this fact. But if we hold fast to extreme holism, everything will seem so frightfully complex that it is doubtful whether we'll be able to do much science at all.

Separating nature into parts seems to be the only way to search for simplicity within nature's complexity. But a position of extreme reductionism might also not allow much science progress, since nature might not be as amenable to separation into parts as this position claims. So although one's position along the realism-idealism range (such as my own hybrid position, presented above) can be justified only *subjectively*, by one's finding it attractive or satisfying, it seems best to determine one's reductionism-holism position by the purely pragmatic criterion of whether it allows, or even encourages, science to progress or not. We now consider three ways by which nature is commonly sliced up by science, and the reductionism-holism position implied by each.

Observer and Observed

Reduction of nature to its simpler parts can be carried out in many different fashions. As the old saying goes, there's more than one way to slice a salami. The most common means of analyzing nature is separating it into two parts: the observer—us—and the observed—the rest of nature. This separation is so obvious that it's often overlooked. It is so obvious because in doing science, in our attempt to understand objectively the reproducible and predictable aspects of nature, we *must* observe nature to find out what's going on and what needs to be understood.

What's happening is this. We and the rest of nature are in interaction; observation is interaction: action and reaction. Anything we observe intrinsically involves ourselves too. The full phenomenon is at least as complex as *Homo sapiens*. Every observation must include the reception of information by our senses, its transmission to our brain, its processing there, its becoming part of our awareness, its comprehension by our consciousness, and so on. We appear to ourselves to be so frightfully complex that we should abandon all hope of understanding anything at all.

So we separate nature into us, on the one hand, and the rest of nature, on the other. The rest of nature, as complex as it might be, is much less complex than all of nature, since our own complexity has been removed from the picture. We then concentrate on attempting to understand the rest of nature. (We also might, and indeed do, try to understand ourselves. But that's another story.) But it's clear that nature with us is not the same as nature without us. (Refer to the discussion of holism in the preceding section.) So what right have we to think that any understanding we achieve by our observations is at all relevant to what's going on in nature when we're *not* observing?

The answer is that, in principle, we have no such right. We are *making an assumption*, or adopting a working hypothesis: The effect of our observations on what we observe is sufficiently weak (or can be made so) that what we actually observe well reflects what would occur without our observation, and the understanding we thus achieve is relevant to the actual situation. This assumption might or might not be valid, its suitability possibly depending on what aspect of nature is being investigated. It is ultimately assessed by the degree of its success in allowing us to understand nature.

The observer-observed analysis of nature is immensely successful in many realms of science. Newton's explanation of Kepler's laws of planetary motion (*An Archetypal Example*, chapter 3) is but one example. This excellent understanding of an aspect of nature was achieved under the assumption that observation of the planets does not substantially affect their motion (a pretty safe assumption, I would say). In general, the separation of nature into observer and observed works well from astronomical phenomena down through ordinary-size phenomena and on down in size to microscopic phenomena.

At the microscopic level, such as in the investigation of individual cells, extraordinary effort must be invested to achieve a good separation. The ever-present danger of the observation's distorting the observed phenomena must be constantly circumvented. That is accomplished in the case of cell research, for example, by the use of microelectrodes and micropipettes, rather than everyday-size devices, for investigating the electrical and chemical aspects of cell behavior.

But at submicroscopic sizes, at the molecular, atomic, and nuclear levels, and at the level of the elementary particles and their structure, the observer-observed separation of nature does not work. It is not a lack of ingenuity or insufficient technical proficiency in designing devices that minimize the effect of the observation on the observed phenomena. Rather, the observer-observed interrelation cannot be disentangled *in principle*; nature absolutely forbids our separating ourselves from the rest of itself.

We look now to quantum theory, which is the branch of science that deals successfully with such matters. From it we learn that nature's observer-observed disentanglement veto is valid for *all* phenomena of *all* sizes. Nevertheless, the *amount* of residual observer-observed involvement, after all efforts have been made to separate, can be characterized more or less by something like atom size. An atom-size discrepancy in the observation of a planet, a house, or even a cell is negligible, whereas such a discrepancy in the observation of an atom or an elementary particle is of major significance.

The observer-observed nonseparability is one aspect of the general nonseparability of nature that quantum theory describes. Nonseparability is not easy to grasp. It is counterintuitive, because we have nothing like it at the level of ordinary-size phenomena. We are used to the idea that, if we have object A here and object B there, then A is clearly here and B is clearly there and they are well separated. But at small sizes we have to be more precise. Object A "being here" means that A is confined to some region of space labeled "here," and similarly for B "being there." A and B being separated means that the two regions of space do not overlap.

But quantum theory teaches us that nature abhors the confinement of submicroscopic objects, such as atoms and elementary particles, to very small regions of space. (That is true for objects of all sizes, but the effect becomes negligible with increasing size.) The smaller the confinement region, the sooner the object will escape and the faster it will move elsewhere! So if we confine A and B, say electrons, to small regions to ensure good separation, they will both soon be anywhere else, and we will have failed to keep them apart. If, on the contrary, we confine them to large regions in the hope they will stay within their assigned regions, the regions will be overlapping, and we will have failed again. And it turns out the same for intermediate-size regions as well. We just can't win! Nonseparability is a general and inescapable fact of nature and cannot be circumvented. (This affect is intimately connected with Werner Heisenberg's [1901–1976] uncertainty principle.)

With regard to the reductionism-holism view of observer-observed separation, the most useful position to hold in the range between the poles—useful in the sense of allowing the best grasp of the situation—is determined by the size of the observed phenomenon: the larger the size, the closer to the reductionism pole; the smaller the size, the closer to the holism pole. The larger the observed phenomenon, the more immune it is to the effects of our observations, whereas the smaller it is, the more susceptible it is to those effects and the more care the observers must take to minimize them. For molecular phenomena and smaller, nearly pure holism seems to be appropriate, since nature's intrinsic nonseparability is highly significant for them.

Quasi-Isolated System and Surroundings

Whenever we separate nature into observer and the rest of nature, we achieve simplification of what is being observed, because instead of observing all of nature, we are then observing only what is left of nature after we have removed ourselves from the picture. Yet even the rest of nature is frightfully complicated. This difficulty might be overcome by a further slicing of nature, by separating out from the rest of nature just that aspect or phenomenon that especially interests us. For example, if we're interested in hydrogen atoms, we might detach one or more hydrogen atoms from the rest of nature and focus our observations on them, while ignoring what is going on around them. Or, in order to study liver cells we might remove a cell from a liver and examine it under a microscope. This further separating of nature was actually tacitly implied in the previous section, where we considered the investigation of phenomena rather than of all of the rest of nature.

But by what right might we think that by separating out a part of nature and confining our investigation to it, while completely ignoring the rest, we will gain meaningful understanding? In principle, we have no right at all, a priori. Ignoring everything going on outside the object of our investigation will be meaningful if the object of our investigation is not affected by what is going on around it. That will be the case if there is no interaction between it and the rest of nature. Such a system, one that does not interact with its surroundings, is called an *isolated system*.

An isolated system is an idealization. By definition, we cannot interact with an isolated system, so no such thing can exist in nature, because nature is, we recall, the material universe with which *we can, or can conceivably, interact*. Well then, how about a system that is observable, that is, is not isolated from *us*, yet *is* isolated from the rest of nature? This is a red herring, since such a system interacts with the rest of nature *through us*. It interacts with us and we interact with the rest of nature, which amounts to an interaction between the system and the rest of nature.

But let's ignore that and look into the question of whether nature allows systems to be isolated from the rest of nature. The state of our present understanding of nature, as incomplete as it might be, is still sufficient to answer the question in the negative.

For instance, no system can be strictly isolated from the gravitational influence of its surroundings. The gravitational force between two bodies decreases in magnitude with increasing separation, but for no separation does the force completely vanish. And no way is known to screen out the gravitational force. So as nearly isolated as a system might be, it will still be subject to gravitational influence from the rest of nature. The best we can do is attempt to minimize this influence by removing the object system as far away as possible from other bodies.

Another kind of nonisolability has to do with *inertia*. Inertia is the property of bodies according to which a body's behavior is governed by Newton's first universal law of motion (*An Archetypal Example*, chapter 3). To wit, in the absence of forces acting on them, or when such forces cancel each other, bodies remain at rest or continue to move uniformly in a straight line. Or expressed positively, inertia is the property of bodies such that force is required to alter their state of motion from one of rest or one of uniform straight-line motion. The state of rest, by the way, is but a special case of uniform straight-line motion: one with constant zero velocity. (Inertia can be made a measurable physical quantity by means of Newton's second law, according to which a body's mass serves as a good measure of its inertia. But this point is not directly relevant to our present discussion.)

Motion that seems uniform and straight-line to one observer might not appear that way to another. For example, if the motion of an aircraft appears to me uniform and straight-line as I watch it while standing on the sidewalk, the same motion could not possibly appear uniform and straight-line to you while you run around me in circles. By Newton's first law, I deduce that all the forces acting on the aircraft cancel each other out completely. But if that is true (and indeed it is), then Newton's first law cannot be valid for you too, since you perceive nonuniform and non-straight-line motion. Still, if you assume Newton's first law is valid for you, then you must deduce that the forces acting on the aircraft do not completely cancel out, and Newton's first law could then not be valid also for me.

Hence it is meaningful to ask: For whom is Newton's first law valid? Relative to what does Newton's first law hold? Or, relative to what is force-free motion uniform and straight-line? And to the best of our present understanding the answer seems to be: relative to all the matter in the universe. Or equivalently, relative to the distant galaxies, where practically all of the matter seems to reside. (And not, strictly speaking, relative to Earth, since Earth undergoes complex motion relative to the distant galaxies (*Reproducibility*, chapter 2). Nevertheless, for many purposes, and certainly for everyday needs, Earth can be used as a sufficiently good reference for Newton's laws.)

That being the case, Ernst Mach (1838–1916) proposed what has come to be known as the *Mach principle:* The origin of inertia lies with all the matter of the universe, that is, the inertia of any body is due to all the other matter present in the universe. Another way of expressing that, in a reductionist manner, is that the inertia of any body is caused by some influence, some interaction, between the body and all the other matter in the universe. About the nature of this Machian influence we know virtually nothing. But since we know of no way to diminish, let alone abolish, the inertia of bodies, it is clear that even for the most nearly isolated systems the Machian influence must remain in full force.

Still, the Mach principle might be grasped better as a holistic principle involving all the matter of the universe, including the body under consideration. Then one would not think of the universe as containing separable bodies with some inertia-causing influence operating among them. Either view, holistic or reductionist, adds an additional anti-isolatory factor to the gravitational influence. (By the way, Einstein was much influenced by the Mach principle when he developed his general theory of relativity.)

As long as the Mach principle is not realized in a manner that science can dig its teeth into, it must be taken as a purely *metaphysical* principle, not a scientific theory. Experimentation so far has revealed little that might indicate the possibility of explaining inertia scientifically. Thus the Mach principle must be considered a guiding principle, offering us a framework for organizing our science concepts, and in no way should it be understood as a theory for the explanation of inertia. At least not for now.

Of the other influences we are aware of, only the forces of electricity and magnetism possess sufficient range to form potential obstacles to the isolation of systems. The properties of these forces are well known. They both weaken with increasing separation, and, in contrast to the force of gravitation, they can both effectively be screened out. That leaves the nuclear forces (the strong and the weak), which are of such short range that they do not hinder isolation.

There does exist an additional anti-isolatory factor, however, although "influence" might not be a good description of it. It is the intrinsic nonseparability of nature as described by quantum theory, which we discussed in the preceding section. Its implication for the isolability of systems is that submicroscopic aspects of systems might be linked to submicroscopic aspects of their surroundings in a holistic manner. The linkage manifests itself through correlations, or interconnections or correspondences, between happenings within the system and happenings outside it. It's not clear whether this linkage can usefully be viewed as some kind of mutual influence or not. In any case, it is uncontrollable, and it can neither be screened out nor be attenuated by large separation.

So the situation with regard to isolated systems is this. To enable themselves to do science, scientists reductionistically try to investigate relatively simple isolated systems. But "isolated" systems are never really totally isolated. They are not totally isolated from us, otherwise we could not observe them (and they would not be part of nature). Neither are they totally isolated from the rest of nature.

The short-range influences—the nuclear forces—are negligible at distances greater than submicroscopic. And the long-range influences—the gravitational, electric, and magnetic forces—can be made as weak as desired by sufficiently removing the system from other matter, and the latter two forces can even be screened out. The putative Machian influence responsible for inertia, if the Mach principle should be viewed this way rather than holistically, cannot be prevented, however, but the properties of inertia are known and can be taken into account. Nevertheless, the quantum correlations that link systems with their surroundings, in manifestation of nature's intrinsic nonseparability, are uncontrollable and unavoidable. They impose an insurmountable limit to isolability and hence also to the usefulness of reductionism, a limit that leaves scientists with no choice but to learn to live with it and to comprehend it holistically.

And those are only the factors we are aware of. The existence of additional, unknown, possibly anti-isolatory factors cannot be precluded. Therefore, we will henceforth use the term *quasi-isolated system* for an "isolated" system, that is, for a system that is as nearly isolated as possible.

Initial State and Law of Evolution

The separation of nature into observer and the rest of nature and into quasi-isolated system and its surroundings are literal applications of the reductionist position. In this section we consider a way of analyzing nature that is a metaphoric application of, or a generalization of, the idea of a part of nature. The two kinds of separation we considered earlier can be envisioned spatially—observer here, observed there, or quasi-isolated system here, its surroundings around it. The analysis in this section is a conceptual separation, the separation of natural processes into initial state and law of evolution.

Things happen. Events occur. Changes take place. *Nature evolves.* That is the relentless march of time. The process of nature's evolution is of special interest to scientists, since predictability, one of the cornerstones of science, has to do with telling what will be in the future, what will evolve in time. Nature's evolution is a complicated process. Yet reproducibility, order, law, and predictability can be found in it, when it is properly sliced. First, the observer should separate himself or herself from the rest of nature (as discussed in the section *Observer and Observed* earlier in this chapter). Then, he or she should narrow the scope of investigation from all of the rest of nature to quasi-isolated systems and investigate the natural evolution only of such systems, as discussed in the preceding section. It is only for quasi-isolated systems that reproducibility, order, law, and predictability are found (*The Lawless Universe*, chapter 5).

Finally, and this is the present point, the natural evolution of quasi-isolated systems should be analyzed in the following manner. First, the evolution process of a system should be considered as a sequence of *states* in time, where a state is the condition of the system at any given time. (This sequence might be continuous or discrete, i.e., smoothly varying in time or jumpy.)

For example, the solar system evolves, as the planets revolve around the Sun and the moons revolve around their respective planets. Imagine that a segment of this evolution is stored on some recording medium. Such a recording is actually a sequence of still pictures. Each still picture can be considered to represent a state of the solar system, the positions of the planets and moons at a particular time. The full recording then represents the recorded segment of the evolution process. (Although the recording happens to be a discrete sequence that jumps from one frame to the next, the actual natural process being recorded is continuous, varying smoothly in time.)

The state of the system at every time should be considered as an *initial state*, a precursor state, from which the following remainder of the sequence develops, from which the subsequent process evolves. For the solar system, the positions of the planets and moons at every time, such as when it is twelve o'clock noon in Rockville on 20 October 2009, say, or any other time, should be considered as an initial state from which the subsequent evolution of the solar system follows.

When that is done, when natural evolution processes of quasi-isolated systems are viewed as sequences of states, where every state is considered as an initial state initiating the system's subsequent evolution, it turns out to be possible to find reproducibility, order, law, and predictability. Here is how that works. Choose what is taken to be a state of the system. In general, there will exist a range of options for that. Among the options, you will find a good choice that will allow you to predict the future. With a good choice of what a state is, you can discover a law that, given *any* initial state, successfully predicts the state that evolves from it at *any* subsequent time. Such a law of nature, since it is specifically concerned with evolution, is referred to also as a *law of evolution*.

Let's return to the solar system. The specification of the positions of all the planets and moons at any single time is insufficient to predict their positions at later times. So the specification of states solely in terms of position is not a good one for the purpose of finding lawful behavior. But the description of states by both the positions and the velocities (i.e., the speeds and directions of motion) of the planets and moons at any particular time does allow the prediction of the state evolving from any initial state at any subsequent time. The law of evolution in this case consists of Newton's three universal laws of motion and law of universal gravitation (*An Archetypal Example*, chapter 3). It operates with much computer number crunching (huge amounts of numerical calculation) and is capable of predicting the positions and velocities of the planets and moons for any time, given their positions and velocities at any earlier time. Astronomers who specialize in the solar system make use of this procedure.

There are even laws of evolution, Newton's laws among them, that work in reverse. For any initial state they not only predict the state that will evolve from it at any later time, but also retrodict the state that occurred at any earlier time and subsequently evolved into the given initial state. (In the case of retrodiction, what we called initial state might more accurately be referred to as final state.) Thus, given the positions and velocities of the planets and moons at any time, Newton's laws allow the calculation of their positions and velocities for any other time, either earlier or later. Newton's laws, moreover, serve as the law of evolution not only for the solar system, but also for any quasi-isolated system of bodies interacting with each other through the gravitational force.

And not only through the gravitational force. This procedure works with electric and magnetic forces as well. Such interaction forces, ones that allow both prediction and retrodiction, are said to possess *time-reversal symmetry*. Not that time can be reversed; it can't. Rather, the processes that result from such interactions can, at least in principle, run in reverse. For the solar system, imagine that at some instant all the planets and their moons reverse their directions of motion. The solar system would then retrace its evolution in reverse; it would de-evolve, one might say. In a thousand years the planets and moons would reach the positions they had in the real solar system a thousand year ago. (Full disclosure: although what I describe is correct, it is an idealization. In the real world the internal structure of the planets and moons affects the evolution, and matters are not as simple as I have made them appear. Similarly, in the regime of the submicroscopic, there exist interaction forces among elementary particles that are not time-reversal symmetric.)

So to discover reproducibility, order, law, and predictability in the natural evolution of quasi-isolated systems, we conceptually split the evolution process into initial state and law of evolution. The usefulness of such a separation depends on the independence of the two "parts": whether, for a given system, the same law of evolution applies equally to any initial state, and whether initial states can be set up with no regard for what will subsequently evolve from them.

Stated in other words, analysis by means of initial state and law of evolution will be useful if these two conditions are fulfilled: on the one hand, nature allows us (at least in principle) complete freedom in setting up initial states, while, on the other hand, what evolves from an initial state is entirely beyond our control. Only then will the analysis be useful, because only then will the two separated "parts" be independent, and only then will we have thereby achieved simplification and found order and law.

The reductionist analysis of evolution processes into initial states and laws of evolution has proved to be hugely successful for ordinary-size quasi-isolated systems, and has served science faithfully for ages. That is how reproducibility is discovered: set up the "same" initial state in the "same" apparatus and obtain the "same" result (*Reproducibility*, chapter 2). It also gives predictability: a law of evolution predicts the outcome of any relevant initial state (*Predictability*, chapter 2). That is the foundation of all laboratory experiments. We rely on this analysis in whatever we do. We insert and turn a key, confident that the door will become unlocked. We move our legs in a certain way and fully expect to find ourselves walking.

The extension of this analysis to the very small seems quite satisfactory, although when quantum theory becomes relevant, the character of an initial state becomes different from what we are familiar with in larger systems. The extension of the analysis to the large, where we cannot actually set up initial states, is also successful (*The Lawless Universe*, chapter 5).

But we run into trouble when we consider the universe *as a whole*, the principal reason being that the concept of law is irrelevant to the universe as a whole (*The Lawless Universe*, chapter 5). Although the universe clearly evolves and at every instant is in some state, in its entirety the universe is in principle not subject to law. (The concept of a beginning, of a literally initial state, is meaningless for the universe as a whole. We don't elaborate on this point.) So, with regard to the possibility of splitting its evolution into initial state and law of evolution, the universe is shaking its head and commanding us to grasp the situation holistically. The states and evolution of the universe as a whole are inseparably intertwined.

Extended Mach Principle

Quasi-isolated systems exhibit reproducibility, order, law, and predictability. And the same laws are obeyed by all systems to which the laws are applicable. As an example, Newton's laws appear to hold for all quasi-isolated astronomical systems of bodies, be they galaxy clusters, single galaxies, binary systems (pairs of stars that are gravitationally closely bound and revolve around each other, a rather common astronomical phenomenon), stars and their planetary systems (like the Sun and the solar system), or planets and their moon systems (like Earth and its single moon or Jupiter and its many moons). So we're approaching the destination toward which we set out at the beginning of this chapter, which is the resolution of the apparent contradiction implied in the question: How is it that within the lawless universe there are laws of nature at all?

But before we can deal with this question, there remains one other station at which we should stop. Before considering the confrontation of the order in the behavior of quasi-isolated systems with the intrinsic orderlessness of the universe as a whole, we should consider the origin of the behavior of quasi-isolated systems. How do quasi-isolated systems "know" how to behave at all? At every moment how does a quasi-isolated system "know" what to do next? How does it constantly obey the *same* laws? How do different quasi-isolated systems obey the *same* laws? How do the system consisting of the planet Jupiter and its moons, on the one hand, and some binary star system located thousands of light-years from us, on the other, both obey the *same* laws (including Newton's laws)?

Let's recall that to find order in a quasi-isolated system, we exploit reproducibility and perform many observations and experiments on many sufficiently similar systems and on the same system itself. Thus the possibility of obtaining data that can be meaningfully compared, and from which order can be extracted, depends on the existence of many systems other than the one being investigated, and on the continued existence in time of the system that is itself under investigation. The other systems serve as reference with which the behavior of the investigated system can be compared, and the continued existence in time of the investigated system allows its behavior (at any time) to be compared with itself (at different times).

That reminds us of the Mach principle (*Quasi-Isolated System and Surroundings*, this chapter), that the origin of the inertia of a body lies with all the other bodies in the universe. There, too, the other bodies are needed to serve as reference for comparing the body's behavior, as reference for its uniform straight-line or other motion. And there, too, the body's continued existence in time is needed to compare its behavior—its motion in the presence and in the absence of forces—with itself. Just as we know of no way to affect the inertia of bodies, so we don't know of any way to change the laws of nature for quasi-isolated systems. All that leads us to enunciate a Machian-like principle, which I called the *extended Mach principle* and which is a generalization of the Mach principle. It is that the origin of the laws of nature for quasi-isolated systems lies with the totality of all systems, that is, with the universe as a whole. Or more succinctly, the origin of the laws of nature is the universe as a whole. (That includes the standard Mach principle as a special case, since inertia is a lawful aspect of nature, as Newton taught us by his first law.)

Additional support for the extended Mach principle, or at least a hint in its direction, is offered by the following reasoning. Let's assume that in some sense it's meaningful to consider the laws of nature for quasi-isolated systems as possessing an origin, and let's inquire into the location of the origin. Could it be within each quasi-isolated system itself? That's not reasonable, since the same laws of nature are found for different systems. Could the origin of the laws of nature be the immediate surroundings of each quasi-isolated system, say within the volume of a room or so? That's ruled out, because the same laws of nature are found in different surroundings. Could the origin be our local region of space, say the solar system or our galaxy, of which the solar system is a constituent? That's not reasonable either, since astronomical observations seem to indicate, or at least are consistent with the assumption, that the laws of nature are the same everywhere and under varied conditions. Thus we're drawn to the whole universe as the origin of the laws of nature.

Yet another line of suggestive reasoning is this. We know that the same laws of nature are valid, wherever relevant, for all quasi-isolated systems. Or at least, as mentioned in the preceding paragraph, our observations are consistent with that. We might find some indication about the origin of the laws of nature by asking what else quasi-isolated systems have in common (besides their being quasi-isolated, which is a prerequisite for their exhibiting laws of nature to begin with). The only answer I can come up with is that they are all part of the same universe, which suggests the extended Mach principle.

(As an aside, tying together various subjects discussed earlier in the chapter, we might note that this answer is that of a realist, whereas an idealist would say that all those quasi-isolated systems are being observed by *Homo sapiens*. From the standpoint of the hybrid position presented in the section *Realism and Idealism* earlier in this chapter, we could say that the origin of the order that is common to all quasi-isolated systems is the whole universe, whereas the origin of the laws that are formed from the order is we ourselves. But let's not unduly complicate matters with these considerations.)

The extended Mach principle, just as in the case of the standard Mach principle, must not be understood as a theory explaining the laws of nature. It is rather a metaphysical guiding principle, offering us a framework for organizing our science concepts. And, as with the standard principle, the extended one might, on the one hand, be viewed reductionistically as indicating the existence of some influence by which the whole universe imposes behavior on small parts of itself. Then we might try to grasp it scientifically by performing experiments to search for and study such an influence. On the other hand, the extended Mach principle might be viewed holistically. Then it would be inappropriate to consider quasi-isolated systems as being meaningfully separable from the rest of the universe, and the behavior of such systems would result unanalyzably from their condition of forming parts of the whole universe.

According to the standard Mach principle, if the rest of the universe were metaphysically and hypothetically taken away (while in some way leaving an uninfluential observer), a lone body would have no inertia. By Newton's first law, an important aspect of inertia is that in the absence of forces acting on it, a body remains at rest or continues to move uniformly in a straight line. Now, a lone body has no forces acting on it, since there is nothing around to exert forces on it. (The uninfluential observer doesn't count.) And a lone body could not have motion. That is because there is no reference against which its motion could have any meaning. (And for the observer, who, being uninfluential, cannot serve as a reference in this scenario, there would be no way of determining whether the body is moving or not.) No motion, no inertia.

Similarly, according to the extended Mach principle, if the rest of the universe were taken away from a quasi-isolated system, not only would the system have no inertia, but all laws of nature would cease to hold for it. There would then be no reference against which anything about its behavior could have any meaning, nothing with which its evolution could be compared. Is the system moving? We just saw this is a meaningless question. Is the system expanding? With no external standard of length, this too is a meaningless question. Is the system even evolving? With no external standard of time, change has no meaning.

The quasi-isolated system would then be truly isolated, and would constitute a whole universe in itself. Science cannot tell us about the fate of such a system, since, as we saw in chapter 5, science is incapable of framing laws of behavior for the universe and, all the more, for universes. If one would like to speculate metaphysically about its fate, one is welcome to do so. But there is no compelling reason to assume that our laws of nature would remain valid for it. The extended Mach principle says they would not. In any case, removing the rest of the universe already belongs to the domain of metaphysics, as do the Mach principle and its extension. And as long as we are involved in metaphysical deliberations, I might append the stinger, which is that there is no compelling reason to think that even space and time would be relevant to such a system.

In summary, then, what do we know of the origin of the laws of nature? Practically nothing, if by "know" we mean "understand through science." It is not even clear whether the concept of origin is scientifically meaningful for the laws of nature. But that should not be surprising, since an origin for the laws of nature would appear to be so fundamental a concept that it very well might not be comprehensible by science. (See chapter 3.) So we philosophize, and our metaphysical considerations lead us to the extended Mach principle, that the origin of the laws of nature is the whole universe. The extended Mach principle can serve as a guide in humankind's scientific endeavor, just as the standard Mach principle has served. Einstein was greatly influenced by the Mach principle, when he developed the general theory of relativity. (See *Quasi-Isolated System and Surroundings* in this chapter, so no déjà vu.)

Whence Order?

It is time to consider the apparent contradiction in the orderly behavior of quasi-isolated systems within the intrinsically orderless universe, in the existence of laws of nature within the lawless universe. The situation is worse now than what we had seen earlier in the chapter. There, the contradiction was apparent when the orderly behavior of a part of the universe was viewed as part of the orderless behavior of the whole universe. But along the way we learned about the extended Mach principle, by which the orderly behavior of a part of the universe is considered to be not only *part* of the orderless behavior of the whole universe, but *brought about* by it. In other words, what previously looked like a contradiction is now a paradox: How can orderlessness bring about order? How can intrinsic lawlessness give rise to laws of nature?

But after all that buildup, what do we do? We take what might appear to be a sneaky tack, a weaseling out. We answer one question with another: Why not? There is no logical imperative that orderlessness bring about only orderlessness, that lawlessness give rise only to lawlessness. Why should orderlessness not have orderly aspects, if only to an approximation (although it doesn't have to)? And why shouldn't lawlessness possess lawful aspects, if only approximately (while, again, it doesn't have to)?

We're already familiar with such happenings, albeit in other situations, and we now consider three such examples. But please don't look for strict analogy with the main object of our investigation. Whereas the orderless universe is considered to bring about orderly behavior of parts of itself, in each of our examples the orderly behavior of a compound system is brought about by the combined orderless behaviors of its parts. The point being exemplified is that order *can* arise from orderlessness, and law from lawlessness.

Consider the human population of some country. Each individual behaves in a way that in many respects is far from orderly. He or she makes or loses money in an unpredictable manner; drives a vehicle erratically; produces babies or refrains from producing them by whim and chance; and falls sick, recovers, and dies quite randomly. Yet the experts who are involved with such matters can predict with some degree of accuracy for the *whole* population such things as birth rate, death rate, average hospital occupancy, savings rate, traffic accident rate, annual total of workdays lost due to illness, etc.

For another example, consider a gas in a closed container, such as air in an otherwise empty corked wine bottle. Because of the immense number of gas molecules in the container under ordinary conditions (roughly on the order of a thousand million million million), the behavior of each individual molecule as it flies around, colliding with other molecules and bouncing off the walls of the container, is unpredictable in practice. Still, properties of the *whole* gas that result from the combined behaviors of its molecules, such as its temperature, its volume, and its pressure on the container walls, do obey orderly relations. For example, if the gas is compressed to half its volume while its temperature is kept constant, its pressure will double.

As our third example, a radioactive isotope, such as C^{14} (carbon-14, an isotope of carbon, an important use of which is the determination of the age of archeological finds), is an isotope (a form of a chemical element) whose atoms' nuclei are unstable and tend to "decay" (to use conventional jargon) by rearranging their internal structure while ejecting one or more subnuclear particles. It is impossible to predict when any individual nucleus will decay. (Quantum theory tells us that is a matter of principle, not due to our technical inadequacy.)

Nevertheless, a whole sample of a radioactive isotope, containing an immense number of atoms, obeys a simple decay law: During equal time intervals, equal fractions (percentages) of the original sample will decay. Thus, if half the sample decays, say, in a year, then during the next year half of the undecayed sample, that is, half of the remaining half, or a quarter of the original sample, will decay, leaving a quarter of the original sample. During the third year a half of that quarter, or an eighth, of the original sample will decay, leaving an eighth. And so on. (By the way, the *half-life* of a radioactive isotope is, as in our example, the time required for the decay of half of any sample of the isotope.)

So we see that orderlessness *can* bring about order. In the examples, that happens by means of an averaging out of the orderlessness of the constituents, resulting in certain orderly behavior of the total system. For the universe as a whole, however, it is the total that is orderless and yet is supposed to bring about orderly behavior of parts of itself. I would be overjoyed if only I could present a good example of that. But I haven't yet found one and welcome suggestions.

The idea of order from orderlessness and law from lawlessness, then, is not so strange after all. And what we have done in this chapter is use the sneaky rhetorical and didactic trick of setting up a straw man only to knock it down. Since we do find in nature order, predictability, and law (for quasi-isolated systems), it follows that the intrinsically orderless, unpredictable, lawless universe possesses orderly, predictable, and lawful aspects. That need not be *precisely* true, however, but must at least be valid to a sufficiently good approximation over a time span containing the period of time humans have been making observations, and for a region of space containing us and the range of our astronomical observations.

Whether nature holds surprises for us over longer time spans or over larger spatial regions is a great unknown. It certainly is possible in principle that the order, predictability, and laws we have discovered will be found to break down in the future, or will be found invalid in newly investigated regions of space as we extend our observational reach. But such speculations are beyond science. (See chapter 4.)

So the order, predictability, and laws we find in nature are among the (approximately) orderly, predictable, and lawful aspects of the behavior of the universe. What about the orderless, unpredictable, and lawless aspects of the behavior of the universe? Do we find any of that in nature? Does nature have its "dark" sides, phenomena that appear to lie outside the framework of science?

Indeed it does! An example on the astronomical scale is that, as the range of astronomical telescopes has been increasing, it has never been found possible to predict just *where* in space the next galaxy, galaxy cluster, or supercluster will be discovered. In other words, the locations of galaxies and their clusters and superclusters in space appear to be an orderless, unpredictable, lawless aspect of nature. (Not totally orderless, however, since it appears that superclusters are concentrated on "walls" in space, with cosmic voids separating the "walls.") Truly cosmic-scale unpredictability, however, is undetectable by us, because the time spans characteristic of the evolution of the universe are immensely longer than the length of time humans have been making scientific observations.

On the scale of the very small, we have an extremely important example of unpredictability in nature: It appears to be impossible to predict individual submicroscopic events. (That was mentioned earlier, in the third example of order from orderlessness, in connection with the decay of nuclei of radioactive isotopes.) This unpredictability holds for submicroscopic events of all kinds, such as the absorption and emission of light by atoms and molecules, the chemical bonding and dissociation of atoms and molecules, nuclear fission (the decomposition of a heavy nucleus into lighter nuclei), and nuclear fusion (the merging of light nuclei to form a heavier one). According to quantum theory, as mentioned earlier, this unpredictability is a matter of principle, and not merely a technical obstacle eventually to be overcome.

On the human scale the situation isn't clear. Some claim the occurrence of phenomena that might be manifestations of the orderlessness of the universe as a whole. These phenomena go under names like anomalous events, transient phenomena, miracles, and parapsychology, where the latter includes effects such as extrasensory perception (ESP), telepathy, telekinesis, and clairvoyance. (We discussed them briefly in *The Lawless Universe* section, chapter 5.) I'm willing to be open-minded about this, but I have not experienced such a phenomenon myself. If these claims are valid, and if at least some of the kinds of reported phenomena are real, yet prove as irreproducible and unpredictable as they are claimed to be, it would seem that we do have human-scale orderlessness and lawlessness in nature.

It shouldn't surprise us that nature possesses an unpredictable side. That's only to be expected, since the universe as a whole is intrinsically unpredictable. We might think it remarkable that nature possesses any *predictable* side at all! We have seen that it does, since we do discover laws of nature. And we have seen there is nothing paradoxical or contradictory in that. But *why* that is so is a valid question. Why does the orderless universe have (approximately) orderly aspects? This is a deep question. Here we aren't looking for an explanation of merely this law of nature or that (see chapter 3), but rather of the very existence of laws of nature. Why is there order in nature?

A satisfactory scientific explanation, that is, a satisfactory theory, of the existence of the laws of nature should entail something more fundamental than the existence of the laws of nature and also more fundamental than the universe as a whole, which, by the extended Mach principle (see the preceding section), engenders the laws of nature (*Generality and Fundamentality*, chapter 3). That's a tall order! The only natural thing I can think of that is conventionally more fundamental than the existence of the laws of nature is the universe as a whole. And nothing at all is conventionally more fundamental than the universe as a whole, since it is, by definition, everything. Thus we find ourselves at the end of science's explanatory power. The existence of laws of nature within the lawless universe appears to be unexplainable by anything that is conventionally more fundamental than itself.

Yet the concept of fundamentality lies outside the strict limits of nature and belongs rather to the domain of metaphysics (*Science and Metaphysics*, chapter 4). Nature does not impose this concept on us, and fundamentality legitimately depends on one's worldview (*Generality and Fundamentality*, chapter 3). There exists within science an unconventional approach called the anthropic principle, which holds that in a certain sense the existence of human beings is more fundamental than the universe as a whole. This point of view is as valid as the conventional one, that there is nothing more fundamental than the universe as a whole. Yet the anthropic principle allows a scientific explanation of the existence of order within the orderless universe, whereas the conventional viewpoint of fundamentality does not, as we just saw. (The anthropic explanation is worked out in detail in the following chapter.)

ℬ

This chapter addresses the question: How is it that within our lawless universe there exist laws of nature?

The metaphysical position of realism is that the order and laws of nature that we find are really "out there," objective and independent of observers. Idealism, however, holds that order and laws of nature are wholly subjective, in the mind of the observer. A possible hybrid position is that order is an objective property of nature, whereas laws are mental constructs.

Holism is the worldview that nature can be understood only in its wholeness or not at all. The opposite position of reductionism is that nature is understandable as the sum of its parts and should be studied by both analysis and synthesis.

One way that science reduces nature to its parts is by the observer-observed separation. It succeeds down to the limit set by nature's intrinsic quantum nonseparability, that is, for phenomena that are not too small. Another way of reducing nature to parts is the separation into quasi-isolated systems and their surroundings, where a quasi-isolated system is one that is isolated from its surroundings to the best of our ability and understanding, albeit imperfectly, owing to inertia and the uncontrollable quantum correlations involved in nature's nonseparability. Nature's order is manifested in quasi-isolated systems, and it is for them that laws of nature are found. A third way that science reduces nature is the analysis of the evolution of quasi-isolated systems into initial state and law of evolution.

The Mach principle, that the origin of inertia lies with all the matter in the universe, is generalized to the extended Mach principle, that the origin of the laws of nature for quasi-isolated systems lies with the universe as a whole. Thus the intrinsically orderless universe possesses (approximately) orderly, lawful aspects, which are the order, predictability, and laws we find in nature. We find evidence also of nature's orderlessness on various scales, including principally the submicroscopic scale, as described by quantum theory. Science can offer no conventional explanation for the existence of laws of nature within the lawless universe.

Bibliography

For realism-idealism, see

R. K. Adair, *The Great Design: Particles, Fields, and Creation* (Oxford University Press, Oxford, 1987).

B. d'Espagnat, *In Search of Reality* (Springer-Verlag, New York, 1983).

M. Eigen and R. Winkler, *Laws of the Game: How the Principles of Nature Govern Chance* (Knopf, New York, 1981, and Princeton University Press, Princeton, N.J., 1993).

H. Fritzsch, *The Creation of Matter: The Universe from Beginning to End* (Basic Books, New York, 1984).

E. Harrison, *Masks of the Universe: Changing Ideas on the Nature of the Cosmos*, 2nd ed. (Cambridge University Press, Cambridge, 2003).

K. Lorenz, *Behind the Mirror: A Search for a Natural History of Human Knowledge* (Harcourt Brace Jovanovich, New York, 1977).

R. Morris, *The Nature of Reality* (McGraw-Hill, New York, 1987).

———, *Dismantling the Universe: The Nature of Scientific Discovery* (Simon and Schuster, New York, 1983).

D. Park, *The How and the Why* (Princeton University Press, Princeton, N.J., 1988).

A. Rae, *Quantum Physics: Illusion or Reality* (Cambridge University Press, Cambridge, 1986).

J. S. Trefil, *Reading the Mind of God: In Search of the Principle of Universality* (Charles Scribner's Sons, New York, 1989).

A. Zee, *Fearful Symmetry: The Search for Beauty in Modern Physics* (Macmillan, New York, 1986, and Princeton University Press, Princeton, N.J., 2007).

For reductionism-holism, see

P. C. W. Davies, *God and the New Physics* (Simon and Schuster, New York, 1983).

J. Monod, *Chance and Necessity: An Essay on the Natural Philosophy of Modern Biology* (Knopf, New York, 1971).

J. S. Trefil, *The Moment of Creation: Big Bang Physics from Before the First Millisecond to the Present Universe* (Charles Scribner's Sons, New York, 1983).

For observer-observed, see Adair (above).

For quantum nonseparability (an anti-isolatory factor), see

J. D. Barrow, *The World within the World* (Oxford University Press, Oxford, 1988).

P. C. W. Davies, *Other Worlds: A Portrait of Nature in Rebellion; Space, Superspace and the Quantum Universe* (Simon and Schuster, New York, 1980).

P. C. W. Davies and J. R. Brown, eds., *The Ghost in the Atom: A Discussion of the Mysteries of Quantum Physics* (Cambridge University Press, Cambridge, 1986).

H. R. Pagels, *The Cosmic Code: Quantum Physics as the Language of Nature* (Simon and Schuster, New York, 1982).

J. C. Polkinghorne, *The Quantum World* (Princeton University Press, Princeton, N.J., 1984).

F. Rohrlich, *From Paradox to Reality: Our New Concepts of the Physical World* (Cambridge University Press, Cambridge, 1987).

See also Adair, Davies (1983), and Rae, above.

For the forces of nature (more anti-isolatory factors), see

F. Close, *The Cosmic Onion: Quarks and the Nature of the Universe* (American Institute of Physics, New York, 1983).

P. C. W. Davies, *The Forces of Nature,* 2nd ed. (Cambridge University Press, Cambridge, 1986).

———, *The Accidental Universe* (Cambridge University Press, Cambridge, 1982).

S. W. Hawking, *A Brief History of Time: From the Big Bang to Black Holes* (Bantam, London, 1988).

See also Adair, Morris (1987), Pagels, and Trefil (1983), above.

For the Mach principle (yet another anti-isolatory factor), see

J. Silk, *The Big Bang,* revised and updated ed. (Freeman, San Francisco, 1989).

See also Adair, above.

For laws of nature, see

J. D. Barrow and J. Silk, *The Left Hand of Creation: The Origin and Evolution of the Expanding Universe* (Heinemann, London, 1983).

R. P. Feynman, *The Character of Physical Law* (MIT Press, Cambridge, Mass., 1965).

V. J. Stenger, *The Comprehensible Cosmos: Where Do the Laws of Physics Come from?* (Prometheus, Amherst, N.Y., 2006).

See also Barrow, Pagels, and Park, above.

In the section *Whence Order?* of this chapter I stated that I would be overjoyed to have a good example of the orderless universe bringing about orderly behavior in parts of itself. I still don't have one. But there is a phenomenon that just might serve as an example, although not one I'm happy about. It is the self-organization of systems. What happens is that certain systems can spontaneously generate and maintain local order within a generally disordered environment. For more details, see

I. Prigogine and I. Stengers, *Order out of Chaos: Man's New Dialogue with Nature* (Bantam Books, New York, 1984).

See also Barrow, above.

For quantum unpredictability, see

T. Hey and P. Walters, *The Quantum Universe* (Cambridge University Press, Cambridge, 1987).

See also Adair, Barrow, Davies (1983, 1980), Davies and Brown, Feynman, Fritzsch, Hawking, Pagels, Park, Polkinghorne, Rae, and Rohrlich, in the lists above.

Some books on irreproducible and unpredictable phenomena are

H. Broch, *Exposed! Ouija, Firewalking, and Other Gibberish* (Johns Hopkins University Press, Baltimore, 2009).

G. Charpak and H. Broch, *Debunked! ESP, Telekinesis, and Other Pseudoscience* (Johns Hopkins University Press, Baltimore, 2004).

M. Shermer, *Why People Believe Weird Things: Pseudoscience, Superstition, and Other Confusions of Our Time* (Freeman, San Francisco, 1997).

CHAPTER SEVEN

Facing the Universe

Human Science

RECALL THE DEFINITION of science as *our* attempt to understand objectively the reproducible and predictable aspects of nature. For this chapter, I strongly emphasize "our." In *Science*, chapter 2, we saw that this seemingly innocuous qualifier carries a heavy load of implication. It tells us that the source of science is within *ourselves*, that science, although having to do with nature, is a *human* endeavor. *Nature*, presumably, would go its merry way whether we were around or not, and whether we tried to understand it or not. But without *our* curiosity and urge to understand, *science* would not exist.

In the view of idealism (*Realism and Idealism*, chapter 6), the essential involvement of science with human beings is obvious and merely a matter of definition. The reproducibility, order, predictability, etc. that we find in nature are, in this view, mental constructs. But even from the viewpoint of extreme realism it should be clear that science is a human endeavor. To do science, we must form a conception of nature. And it must be admitted that, as real as realists would have nature and its reproducibility, order, and so on be, our conception of all that is human-dependent. It is formed only after filtering by the interactions between *us* and the rest of the universe (*our* observations and measurements), filtering by *our* senses, processing by *our* nervous system, *our* conscious awareness of *our* perceptions, and processing of *our* perceptions by *our* consciousness. Then *we* try to understand what *we* conceive, and *we* consider it understood when *we* have found explanations that satisfy *us*.

Dolphins, for example, probably have some kind of dolphin science, which would be *their* attempt to understand nature as *they* conceive it. *Their* conception of the universe is formed out of *their* perceptions, which are obtained through *their* senses and nervous system, and processed by *their* consciousness. Dolphins' conception of the universe is in all likelihood quite different from ours, since their senses are different from ours, as is certainly their consciousness as well. *Their* science would involve the devising of explanations that are satisfying to *them*. It's hard to know what kinds of explanation are satisfying to dolphins, but in sum we can reasonably assume that their science is different from ours. Yet dolphin science is as valid for them as ours is for us.

Thus *Homo sapiens* plays an essential role in (human) science in the sense that: (1) science is a by-product of *our* existence; (2) it is *our* conception of nature that we attempt to explain; and (3) a valid explanation is one that satisfies *us*. Point 1 is obvious. Point 2 is nearly obvious: conception of nature can well be assumed to be species-dependent, and it's surely not dolphins' or dogs' conception that we attempt to explain. (Even among people, conception of nature is known to be culture-dependent. Thus science should be, and is, different in different cultures as well as in the same culture at different times, as the culture evolves. To the extent we are approaching something like a world culture, at least with regard to a conception of nature, we can speak of a "universal" science. But we leave that issue aside.) Point 3 is also nearly obvious, especially after our extensive discussion in chapter 3. Nature imposes no criteria of acceptability for theories; a theory is acceptable when it satisfies *our* feeling that something is indeed being explained.

Anthropic Principle

In addition to our essential role in science, that is, in our attempt to understand nature objectively, we also have a role in the proceedings of nature itself, as we humans are definitely part of the material universe. Since our existence is a natural phenomenon, our curiosity drives us to try to understand it within the framework of science, to explain it by other aspects of nature. Darwin's theory of biological evolution and its modern versions are such attempts. (Holders of transcendent worldviews, however, might try to explain our existence outside the framework of science. Creationism and "intelligent design" are examples of such attempts.)

Moreover, to invert matters, the existence of human beings, as a natural phenomenon, can in principle serve as an explanation for other natural phenomena. For example, our existence explains much of the environmental pollution on this planet. And our existence might possibly explain other, more general and more fundamental aspects of nature. The *anthropic principle* states, for a start, that the existence of *Homo sapiens* may, within the framework of science, serve as an explanation for phenomena and aspects of nature. (In the following we'll make an important addition to this statement.)

At this point it is useful to reiterate: We're taking the term "understand" to mean "be able to explain." By "explain" we mean "give reasons for." An explanation is a reason (or a set of reasons) for something. (See *Science*, chapter 2.) A theory is a scientific explanation (see chapter 3).

Consider the following. Astronomical data, as interpreted according to the cosmological schemes currently in vogue, indicate the age of the universe to be about 14 thousand million years. Why is the age of the universe just this and not otherwise? If it were much *less*, by the same cosmological schemes, there would not have been sufficient time for the production (in stars) of the heavier chemical elements necessary for our existence. And if it were much *more*, all the stars would have burned out and we could not survive. Therefore, we shouldn't be surprised to find the age of the universe to be what it appears to be.

Note how the question about the age of the universe is answered in terms of why we can't be existing and observing it in any era but the present one. Our existence is used to explain, to give a reason for, the age of the universe we discover through our observations.

That is an application of the anthropic principle. Admittedly, the age of the universe, which is what is being explained here, does seem to be a rather incidental aspect of nature, since it is changing all the time. For an anthropic explanation of something apparently more fundamental, consider the following.

The anthropic principle has been invoked, among other applications, to explain the strength of the gravitational force, for which no conventional, that is, nonanthropic, explanation is presently known. Without going into technical details, if the strength of the gravitational force were much different from what it is, stars would evolve differently from the way they have been, and are, evolving. Their evolution would not allow the formation of planets, nor the development of planet-dwelling observers such as ourselves. Hence our existence implies that the gravitational force has the strength it does and none other; our existence is a logically sufficient condition for the gravitational force to have the strength it does. That is a valid reason, and thus our existence can be taken as explaining the strength of the gravitational force. A similar explanation can be made also for the strength of the nuclear force that binds protons and neutrons together to form atomic nuclei.

We're doing some legitimate logical manipulation here. If you're unfamiliar with it, or a bit rusty, let's briefly review the idea. Our reasoning is based on the following. If the nonexistence of B implies the nonexistence of A, then, and completely equivalently, the existence of A implies the existence of B. For example, "No clouds in the sky implies no rain" is equivalent to "Rain implies clouds in the sky." Or another: The unavailability of beverage implies that no drinking is going on. That is equivalent to the act of drinking implying that there is beverage available.

Let's try another for practice. If you're not sleeping, you can't be dreaming. Thus if you're dreaming, you must be sleeping. And now, in the above anthropic explanation, "The gravitational force not having its observed strength implies our nonexistence" is equivalent to "Our existence implies the observed strength for the gravitational force."

Some terminology: When the existence of A implies the existence of B, then the former is termed a *sufficient condition* for the latter, while the latter is called a *necessary condition* for the former, and the latter can be said to *logically follow from* the former. For example, rain is a sufficient condition for clouds in the sky, clouds in the sky are a necessary condition for rain, and there being clouds in the sky logically follows from there being rain. Similarly, the act of drinking is a sufficient condition for the availability of beverage, the availability of beverage is a necessary condition for drinking, and the availability of beverage logically follows from the act of drinking.

Also: When the nonexistence of B implies the nonexistence of A, then the nonexistence of B is a valid reason for the nonexistence of A and thus explains it. Conversely and equivalently, as we have seen, the existence of A is then a valid reason for the existence of B and explains it. For our rain example, no clouds in the sky is a reason and explanation for no rain. We grasp that both logically and causally. Since rain comes only from clouds (although not all clouds produce rain), the absence of clouds surely precludes rain. The absence of clouds indeed *causes* a lack of rain, so the former also logically implies the latter.

But equivalently, rain can be considered a reason and an explanation for there being clouds in the sky. As we have seen, the presence of clouds certainly follows logically from the presence of rain. To convince ourselves of that, consider this conversation: "There must be clouds in the sky." "How do you know? You can't see the sky." "True, but I can hear the rain."

But here we don't grasp rain as *causing* clouds; we don't see any mechanism whereby rain *brings about* clouds. So although rain is a logical reason and explanation for the presence of clouds, we don't find it a satisfying one.

In like manner, the unavailability of beverage clearly causes there to be no drinking. Thus beverage unavailability is both a logical and a causal explanation and reason for absence of drinking. After all, you need beverage to be able to drink. Equivalently, the act of drinking is reason and explanation for the availability of beverage. It is a logical reason and explanation. Again: "There must have been beverage available last night." "How do you know? I never mentioned beverage." "True, but you told me you were drinking last night."

However, we don't see any mechanism by which drinking *causes* the availability of beverage. So again, although drinking is a logical reason and explanation for the availability of beverage, we don't feel comfortable with it.

Let's return to the gravitational force and our existence. If the strength of the gravitational force were different from what it is, we couldn't exist, assuming the experts know their job. So a different strength *causes* our nonexistence (through stars evolving differently and planets not forming). Thus a different strength would be a logical and causal reason and explanation for our nonexistence. But it's a logical equivalence that our existence is a reason and explanation for the gravitational force having the strength it actually has. That is the anthropic explanation of the strength of the gravitational force. It's clearly an unsatisfactory explanation, because we don't see any causation; our existence doesn't *cause* the gravitational force to have this, or any, strength.

Although not a causal explanation, it *is*, nevertheless, a logical explanation: "The gravitational force must have such and such a strength." "How do you know? You're a theoretician, and if you ever tried to measure the strength of the gravitational force, the apparatus would certainly go irreversibly haywire." "True, but I exist."

Let's use the anthropic explanation of the strength of the gravitational force as a case study to help clarify the anthropic principle. The explanation runs into two difficulties, which we might call to subjective and objective. The subjective difficulty is that scientists, and most likely you too, just don't *feel* that any explaining is being done, since there is no causation, as we just saw. This is the standard objection to the anthropic principle. Recalling chapter 3, we want what is doing the explaining to logically imply what is being explained, to be more general, more fundamental, more unifying, and simpler than what is being explained, and we would also like the former to be the cause of the latter.

Of all these desirable attributes, it seems that only the first (logical implication) is true, the prerequisite of a scientific explanation, without which the explanation would not have been proposed to begin with. As we saw, our existence logically implies the strength of the gravitational force, because, if it were much different, we would not exist. On the other hand, our existence does not at all seem to be more general, more fundamental, more unifying, or simpler than the strength of the gravitational force. Neither do we perceive our existence as causing the latter. If anything, the opposite would appear to hold—the strength of the gravitational force is more general, more fundamental, more unifying, and simpler than our existence, and it is even possible that the strength of the gravitational force is part of the cause of our existence.

We must gracefully concede lack of generality, unification, simplicity, and causation in the anthropic explanation. And that is the subjective difficulty of the principle when it is applied to the explanation of apparently fundamental aspects of nature. Without those, it's hard to feel that an anthropic explanation is acceptable.

We must not concede lack of fundamentality, however. After all, who are we to tell the universe what is more or less fundamental for it? Viewed holistically, the universe is an integrated, all-encompassing whole, of which all aspects and phenomena are interrelated and interdependent. We can't experiment with variations of the universe, so we can't know which of its aspects and phenomena, if any, are more or less fundamental. Thus absolute fundamentality is an empty concept in science. The most we can do is point out what aspects of nature *seem* more or less fundamental *to us* on the basis of *our* understanding of nature.

Fundamentality, therefore, is both relative to our state of understanding and a matter of convention. Having recognized this fact, we can come to the realization that our existence (the existence of each of us, generalized to *Homo sapiens*) is the most fundamental aspect of nature *for us*, because it underlies our perception of other aspects of nature. Some people might not consider this reasoning to be satisfactory, but it is valid.

After all, in the whole of nature, what phenomenon are we most sure of, have the least doubts about, have the most confidence in, if not *our own existence*? Consider what a person deprived of all sensory perception would be aware of. Whereas it is fairly innocuous to doubt any other natural phenomenon, consider the paradox involved in doubting one's own existence. Indeed, cogito ergo sum (I think, therefore I exist)! Add to that the nature of science, as an intrinsically *human* endeavor.

So, although we concede a subjective difficulty for the anthropic explanation in the case under study and in general, the point of fundamentality is *not* part of that difficulty, when considered from the enlightened vantage we just gained. Thus the anthropic principle not only states that the existence of *Homo sapiens* may serve as an explanation for other aspects of nature, but the principle can offer the most fundamental explanations, since it bases them on the most fundamental natural phenomenon we have, our own existence.

We now state the anthropic principle in its full form:

The existence of Homo sapiens *may, within the framework of science, serve as an explanation for phenomena and aspects of nature, and moreover, such explanations are the most fundamental.*

The objective difficulty presented by the anthropic explanation in our case study, and in similar cases, is what I call the invariant-context problem. As we are aware, for an "explanation" to be an explanation, at the very least what is doing the explaining must logically imply what is being explained. In our case, the existence of *Homo sapiens* must logically imply that the strength of the gravitational force is as it actually is, or, the actual strength of the gravitational force must follow logically from our existence, or, if the strength were different we could not exist. (Refer to our logic review earlier in this section.) Now, it has been shown that, if the strength of the gravitational force were much different from what it is, we would not be around to wonder about it. Thus the actual strength does seem to follow from the existence of *Homo sapiens*.

But it is important to be aware of the context of this explanation. While we're considering the hypothetical possibility of varying the strength of the gravitational force, we're tacitly assuming that no other aspect of the universe, no other law of nature, nothing else at all, is being varied. That is the invariant-context problem, because we can never be sure that a change in the strength of the gravitational force cannot be compensated for by some concomitant change in other laws of nature.

This kind of qualification must be kept constantly in mind in all applications of the anthropic principle, even if not stated explicitly. Almost all applications of the anthropic principle assume an invariant context, which qualifies their explanatory power. We can almost never be sure that some concomitant variation in the context will not compensate for the assumed variation in the situation being explained, so that the existence of *Homo sapiens* will not be precluded after all.

So where do we stand with the anthropic principle? Because of the subjective difficulty presented earlier, the anthropic principle should be resorted to only when no conventional explanation is available. In other words, we should us it only for explaining aspects of nature that appear to be so fundamental that we are hard put, or at a loss, to find an aspect of nature that conventionally appears to be even more fundamental. Such an aspect, for example, might be space, time, the laws of nature, or the nonseparability of nature. When we run into such a situation, we can declare, "Aha! We thought that aspect is so fundamental. But we're now anthropically enlightened, and we know our existence is even more fundamental (cogito ergo sum!)." And we might devise an anthropic explanation for so fundamental an aspect of nature.

True, it would not be a satisfying explanation, since it would not meet the criteria of generality, unification, simplicity, or causation. But it would fulfill the criteria of logical implication and fundamentality, and would thus be a viable explanation, where the conventional alternative offers no explanation at all. With anthropic

enlightenment and good will, one can get used to anthropic explanations, especially when no conventional ones are available. Nevertheless, the explanations offered by the anthropic principle are liable
to be qualified by the assumption of invariant context, which is their
objective difficulty.

Explanatory chains can be set up in nature. Say we wish to explain some aspect or phenomenon of nature. We look for another,
conventionally more fundamental aspect or phenomenon of nature
that explains the former. Upon finding it we're back to square one,
because we're not satisfied leaving this aspect unexplained either.
And so it goes, from aspect to conventionally more fundamental
aspect. Until we have gotten so fundamental that we can't find any
conventionally more fundamental aspect to explain our last explaining aspect, except perhaps the universe itself, which amounts to no
explanation at all. At this stage, we recognize that our existence is
really the most fundamental aspect of nature *for us*, and perhaps
succeed in thus explaining the last extremely fundamental aspect we
were trying to explain.

Let's say we want to explain the existence of *Homo sapiens*, which
conventionally appears to be an incidental phenomenon of nature.
We might appeal to Darwinian evolution at the first level of explanation. Then we might explain Darwinian evolution by means of
molecular biology at the next level. The following level might be an
explanation of molecular biology through biochemistry. Then we
might explain the latter by physics, that is, by quantum theory and
the other basic laws of nature. And then what? We have no conventional explanation for quantum theory and the other basic laws of
nature. So rather than end the explanatory chain at this point, we
could invoke the anthropic principle and perhaps explain quantum
theory and the other basic laws of nature by the existence of *Homo
sapiens* as the most fundamental aspect of nature.

What a beautiful circularity: our existence (viewed fundamentally) explaining our existence (viewed incidentally)! Should that
bother us? Not at all. We realize we're not dealing with just any
universe, or with some general model of a universe, but rather with

the only universe we have. And we realize that the universe *contains us in an essential way* by the very fact of *our* investigating it. A relevant concept here is "self-reference": The object of the investigation intrinsically contains its investigators as part of itself; the investigators are investigating an object of which they are intrinsically a part. That is the holistic worldview within which the anthropic principle operates.

Let's make a clarification at this point. I reemphasize that the anthropic principle does *not* imply causation, and an anthropic explanation does *not* mean that the existence of *Homo sapiens* is the cause of whatever it is explaining. This is one of the unsatisfactory properties of anthropic explanations. There is no teleology (no design or purpose) or religion here, and the anthropic principle is no return to anthropocentrism, that is, it does not reinstate humanity to its historically former privileged position as the center of nature. It does, however, recognize our privileged position as the center of science (one of the centers, anyway), since science is, it cannot be overemphasized, a *human* endeavor.

Neither is it claimed that *Homo sapiens* is the reason for the universe. That would be metaphysics, whereas the anthropic principle is a scientific principle. An anthropic explanation is, however, an explanation of last resort, to be used only when no conventional explanation is available, in order to provide, where applicable, some—however unsatisfactory—scientific explanation rather than having none at all.

Additional comments are in order here. As we derived it above, the anthropic principle is this: The existence of *Homo sapiens* may, within the framework of science, serve as an explanation for phenomena and aspects of nature, and moreover, such explanations are the most fundamental. In order to give a more nearly balanced perspective, I should mention that not all scientists accept the anthropic principle as stated here; some object to the part that such explanations are the most fundamental, and some to the whole thing. In addition, other versions of the anthropic principle have been proposed, and they too have their objectors. The term "anthropic principle" covers a variety of principles, all having the common feature of somehow involving the existence of *Homo sapiens*. Support for or objection to any one version is not necessarily valid for or against other versions.

I won't burden you with other versions. But I will add that the anthropic principle we derived above is the only version I know of that both is a *scientific* principle (some versions are metaphysical) and, in my opinion, possesses deep significance. It is concerned not only with what is going on in nature but also with science, *our* attempt to understand objectively what is going on in nature, as a *human* endeavor. It recognizes that we, as investigators of nature, are intrinsically part of the object of our investigation, so that in the final reckoning we can't avoid coming back to ourselves in some manner or other.

We can thus appreciate Arthur Stanley Eddington's (1882–1944) beautiful description, though let's not interpret it too extremely: "We have found a strange footprint on the shores of the unknown. We have devised profound theories, one after another, to account for its origin. At last, we have succeeded in reconstructing the creature that made the footprint. And lo! it is our own."

Whence Order? (Again)

In *Whence Order?*, chapter 6, we raised the question of why there is order within the orderless universe, why, indeed, nature possesses any predictable side at all. There we saw that no conventional answer to this question should be expected, since what we want to explain is too fundamental for conventional explanation. Thus the situation is ripe for an anthropic explanation, if it can be shown that some predictability and order are necessary for our existence.

We can't give a rigorous proof, but we can give a plausibility argument, which might run as follows: Without order and predictability we could not reliably control our bodies or anything else, and could not survive our interactions with our unpredictable environment and with each other. Also, consider the degree of constancy of the laws of nature, both throughout space and over time, that is necessary for our survival as individuals and as the human race, for us to have stable, reliable memories as individuals and as a society, and so on. Because of our physical complexity it is plausible that the bounds on the possible variability of the laws of nature are extremely tight, at least throughout some sufficiently large volume of space and over some sufficiently long period of time. This is so because the complex systems that are us exist *stablely* under the *present* laws of nature. Judging from our experience with complex systems, it is very plausible that this dependence of stable existence on the laws of nature is a critical one, that even small changes in the laws of nature would destabilize us, bringing about our disintegration.

If we take this plausibility argument seriously, the anthropic reasoning (refer to the logic review in the preceding section) then runs: Since lack of order implies our nonexistence, it is logically equivalent that our existence implies order. Thus, since we do exist, the lawless universe must have orderly aspects, and these orderly aspects must, and do, affect us strongly enough to allow our existence. They are the laws of nature we discover. The orderless aspects of the universe must and do affect us only weakly; otherwise, by the same reasoning, we could not survive.

As we saw in *Extended Mach Principle*, Chapter 6, we know about submicroscopic orderlessness, the essential quantum unpredictability on the submicroscopic scale. It causes mutations, maybe cancer, and perhaps other unpleasant results. But although individual personal tragedies may be grave, the effect on the whole is weak, since most people manage to survive and the human race thrives. (Mutations even form a crucial component of biological evolution.) Possible human-scale irreproducibility, if it exists, might show up as "anomalous events," "transient phenomena," miracles, parapsychology, etc., which appear only rarely and affect us only slightly on the whole. (Such effects were discussed briefly in *The Lawless Universe* section, chapter 5). Larger-scale orderlessness appears not to affect us at all.

That is how the anthropic principle explains the existence of laws of nature within the lawless universe. In this case, the explanation is especially unsatisfying, because we can't even show that our existence implies what it is supposed to explain, and so we have to make do with a mere plausibility argument. But the anthropic principle does give us some explanation for what would conventionally remain unexplained.

Space and Time

The conventionally most fundamental aspects of nature are, it seems to me, space, time, and nature's quantum aspect (including submicroscopic unpredictability and uncontrollable correlations). Since there can be nothing conventionally *more* fundamental than *most* fundamental, no *conventional* explanation of space, time, and the quantum is to be expected. It might, however, be possible to explain one or two of them by the other two or one. We might, say, explain space and time by nature's quantum aspect. Some researchers have attempted to do that, but the results have not been greatly convincing. So, as in the preceding two sections, we again have a situation that is ripe for the anthropic principle. Since space and time can't be explained conventionally, we invoke the existence of *Homo sapiens* (considered fundamentally) to explain them. The anthropic principle offers a clean, unqualified explanation of space and time. Here it is.

To begin with, we need working definitions for space and time. As one might well imagine, much has been said and written about that, mostly philosophical. My own preference is this: Space is the dimension of material being; time is the dimension of becoming. These definitions are extremely concise, which is an advantage. But their conciseness tends to hide their complexity, which is the complexity inherent to the concepts of dimension, being, and becoming. The concepts of being and becoming have been the subjects of philosophical deliberations for ages, and these deliberations continue as vigorously as ever. For the purpose of the present discussion, we'll exempt ourselves from the fray and leave the concepts of being and becoming to be understood by you in whatever way you ordinarily understand them.

The concept of dimension, however, is more technical and certainly less familiar. The term has several related meanings, but the one that is relevant to our discussion is that dimension is the possibility of assigning a measure to something. Thus we can speak of a musical tone as possessing the dimension of pitch, measured by frequency (in hertz), the dimension of intensity, measured by power per unit area (in watts per square meter), and the dimension of timbre (or, tone quality), which can be measured also. Or, color is said to possess the dimensions of hue, measured by two "chromaticity coordinates" (numbers between zero and one), of saturation, measured by purity (again a number between zero and one), and of intensity, measured by power per unit area (in watts per square meter). Often, a dimension is assigned to each number involved in a measure. Thus color is said to possess four dimensions: two dimensions of hue (two chromaticity coordinates), one dimension of saturation (purity, a single number), and one of intensity (the number of units of power per unit area). The shape of a packing box has three dimensions: its length, its width, and its depth, where each is measured by linear extent in centimeters, say, or in inches.

Material being, too, allows the assignment of a measure and thus possesses dimension. How do we assign a measure to being? By answering the question "Where?" For example, "Where is it?" How does that work? Material being is what is most fundamentally inherent to every material object, its *isness*, so to speak. There is no material object without being, and no material being without some object having it. (Would you believe this is what some philosophers do all the time?) We can ask and answer the question "Where?" for every material object, so material being (of some object) can be assigned a measure by answering the question "Where?" for the object possessing it. Our experience teaches us that space, which we defined as the dimension of material being, consists of three dimensions, since the question "Where?" always requires three numbers for a full answer: latitude, longitude, and altitude with respect to Earth, for example. Dimension is the possibility of assigning a measure, and here we are assigning a measure. So space is the assignment of a measure involving three numbers to material being, in answer to the question "Where?"

What we have found is that the above definition of space as the dimension of material being is not as weird as it might have appeared at first. It is what we usually think of as space: three directions for designating the location of an object: left-right, front-back, and up-down.

Now for time as the dimension of becoming. This means assigning a measure to becoming. Becoming involves change, change in the state of material being. Thus it presupposes material being. But it goes beyond mere being in that it always allows the assignment of a measure by the answer to the question "When?" "When did (or does or will) the change occur?" for example. So, by the definition of time as the dimension of becoming, time is just this possibility of answering the question "When?" with regard to becoming. From our experience we know that time consists of a single dimension, since the question "When?" can always be answered by a single number, composed of the date and the clock reading, for example. So our definition of time as the dimension of being is what we usually think of as time. Nothing strange about it.

After those preliminaries, we are prepared to use the existence of *Homo sapiens* to explain space and time. The first step is to show that the existence of *Homo sapiens* implies being and becoming. Well, we obviously *are* and we clearly *become*. But to sharpen things a bit, let's extract from "the existence of *Homo sapiens*" just "the existence of a learning system." We are systems that exist and learn. Then: Existing is being, and learning implies becoming (specifically, becoming more learned). Thus the existence of *Homo sapiens* implies being and becoming.

The next step of the anthropic explanation of space and time is to recognize that everything implies its own dimension(s). For instance, color implies its dimensions: hue, saturation, and intensity. A rectangle implies its dimensions: length and width. And similarly, being implies its dimension: space. And becoming implies its dimension: time.

The third step is to combine the first two steps. The existence of *Homo sapiens* implies being, which in turn implies space. Thus our existence implies space. And our existence also implies becoming, which in turn implies time. Thus our existence also implies time.

The final step of the anthropic explanation is to take the logical implication as a reason and thus as an explanation. Hence the existence of *Homo sapiens* explains space and time. It's as simple as that. A clean, unqualified application of the anthropic principle. But an unsatisfactory explanation, you claim? Sure it is, as we discussed earlier in the chapter. Yet, at least for now, it's the only scientific explanation of space and time, of any kind, to be found!

༄

Science is a human endeavor, since *Homo sapiens* plays an essential role in it, in the sense that: (1) science is a by-product of *our* existence; (2) it is *our* conception of nature that we are attempting to explain; and (3) a valid explanation is one that satisfies *us*. The anthropic principle is that the existence of *Homo sapiens* may, within the framework of science, serve as an explanation for phenomena and aspects of nature, and moreover, such explanations are the most fundamental. Fundamentality follows from the fact that our existence is, *for us*, the most fundamental aspect of nature.

Anthropic explanations should be used only as a last resort, for aspects of nature that are apparently so fundamental that no conventional explanation is available. That is because anthropic explanations suffer from the subjective difficulty that our existence seems neither more general, more unifying, nor more simple than whatever it is that's doing the explaining, and we do not perceive our existence as *causing* whatever it is that's, being explained either. Anthropic explanations also suffer from the objective difficulty of the invariant-context problem.

We give the existence of order within the orderless universe a rather weak anthropic explanation. But we give space and time a clean, unqualified anthropic explanation.

Bibliography

For science as a human endeavor, see

R. K. Adair, *The Great Design: Particles, Fields, and Creation* (Oxford University Press, Oxford, 1987).

J. D. Barrow, *The World within the World* (Oxford University Press, Oxford, 1988).

B. d'Espagnat, *In Search of Reality* (Springer-Verlag, New York, 1983).

M. Eigen and R. Winkler, *Laws of the Game: How the Principles of Nature Govern Chance* (Knopf, New York, 1981, and Princeton University Press, Princeton, N.J., 1993).

H. Fritzsch, *The Creation of Matter: The Universe from Beginning to End* (Basic Books, New York, 1984).

E. Harrison, *Masks of the Universe: Changing Ideas on the Nature of the Cosmos*, 2nd ed. (Cambridge University Press, Cambridge, 2003).

K. Lorenz, *Behind the Mirror: A Search for a Natural History of Human Knowledge* (Harcourt Brace Jovanovich, New York, 1977). (This book is especially concerned with the evolutionary adaptation of our cognitive facility to the reality of our environment.)

J. Monod, *Chance and Necessity: An Essay on the Natural Philosophy of Modern Biology* (Knopf, New York, 1971).

R. Morris, *The Nature of Reality* (McGraw-Hill, New York, 1987).

————, *Dismantling the Universe: The Nature of Scientific Discovery* (Simon and Schuster, New York, 1983).

D. Park, *The How and the Why* (Princeton University Press, Princeton, N.J., 1988).

I. Prigogine and I. Stengers, *Order out of Chaos: Man's New Dialogue with Nature* (Bantam Books, New York, 1984).

F. Rohrlich, *From Paradox to Reality: Our New Concepts of the Physical World* (Cambridge University Press, Cambridge, 1987).

For discussions, applications, and other versions of the anthropic principle, see

J. D. Barrow and J. Silk, *The Left Hand of Creation: The Origin and Evolution of the Expanding Universe* (Heinemann, London, 1983).

J. D. Barrow and F. J. Tipler, *The Anthropic Cosmological Principle* (Oxford University Press, Oxford, 1986).

P. C. W. Davies, *The Accidental Universe* (Cambridge University Press, Cambridge, 1982).

————, *Other Worlds: A Portrait of Nature in Rebellion; Space, Superspace and the Quantum Universe* (Simon and Schuster, New York, 1980).

————, *Space and Time in the Modern Universe* (Cambridge University Press, Cambridge, 1977).

J. Gribbin and M. Rees, *The Stuff of the Universe: Dark Matter, Mankind and the Coincidences of Cosmology* (Heinemann, London, 1989).

S. W. Hawking, *A Brief History of Time: From the Big Bang to Black Holes* (Bantam, London, 1988).

See also Adair and Barrow, above.

Some books on irreproducible and unpredictable phenomena are

H. Broch, *Exposed! Ouija, Firewalking, and Other Gibberish* (Johns Hopkins University Press, Baltimore, 2009).

G. Charpak and H. Broch, *Debunked! ESP, Telekinesis, and Other Pseudoscience* (Johns Hopkins University Press, Baltimore, 2004).

M. Shermer, *Why People Believe Weird Things: Pseudoscience, Superstition, and Other Confusions of Our Time* (Freeman, San Francisco, 1997).

CHAPTER EIGHT

The Hunt for Reality

THE NATURE OF reality has been the subject of extensive, deeply searching philosophical thought, discussion, and argument throughout the ages. In our own age the issue is still far from resolved, and arouses undiminished fervor. As science has developed from, say, the 1600s to the present, it has taught humankind objective facts about reality that thinkers cannot afford to ignore. And especially in our own centuries, the implications of the attainments of science, as expressed in relativity and quantum theories, are crucial for our understanding of reality. So much so, I feel strongly that any educational curriculum worthy of its name must assure some minimal elementary understanding of these theories.

Even so, I repeat my assertion that considerations about reality are philosophical considerations, even metaphysical, in the sense described in *Science and Metaphysics*, chapter 4, since they bear strongly on science. Science can constrain metaphysical speculations about reality, but it cannot totally replace them. Science can go only so far in telling us about reality; reality's ultimate nature is beyond the domain of science. Yet it seems to me that science, although it cannot force one to adopt any particular worldview, strongly hints in a certain direction. If we're willing to let science guide us, we will be led to a particular worldview.

Metaphysical Positions

First, let's reconsider the metaphysical position of *realism*, presented in *Realism and Idealism*, chapter 6. There, realism was presented as the position that the order and laws of nature are inherent to the world we observe, and are independent of observers. I opposed realism to *idealism*, in which the order and laws of nature exist wholly in the mind of the observer. I then proposed a hybrid position, that nature does possess objective order, whereas laws are a human device.

Realism, in a more fundamental sense of the term, is the metaphysical position that there is an objective, observer-independent underlying reality that we discover and study through our physical senses (and through our measuring and observing instruments as extensions of our senses). Realism claims that there is a real world "out there" existing independently of us. The assumption of realism in this sense, made in chapter 1, pervades the presentation in this book, simply because my own worldview is basically realist, as is that of almost all scientists. Realism in this sense is usually placed in opposition to *positivism*, which is the metaphysical position that only our sense data, derived from measurements and observations, are fundamental. Positivist science is thus expressed solely in terms of measurements and observations, solely in terms of the phenomena themselves. It refrains from considerations of any reality that underlies the phenomena, any reality that would not be directly accessible by our physical senses and their extensions.

Realists and positivists both agree that our measurements and observations are the source of our scientific knowledge of nature. Positivists claim that it is meaningless to go beyond these, that nothing significant can be said about whatever might be underlying the phenomena. Realism goes beyond positivism by assuming that an objective reality underlies natural phenomena, and is the ultimate source of the results of our measurements and observations. Thus, according to the realist view, our measurements and observations, as interesting as they might be in themselves, tell us something about objective underlying reality.

As an example of the difference between realist science and positivist science, consider a hypothetical investigation of a sample of some radioactive material. Assume that one gram of the material is placed two centimeters away from the window of a Geiger counter.

The counter responds with a continual irregular series of clicks, while a pointer indicates the average number of clicks per minute, the count rate. (The technical details aren't important, but I give them to make the procedure sound authentic.)

Both realist and positivist scientists would be interested in the dependence of the count rate on the distance between the sample and the counter window, as well as its dependence on the thickness and kind of material placed between the sample and the counter window. Both scientists would be interested, as well, in the dependence of the count rate on time. Both would make many measurements of many kinds, analyze the results, and look for order.

Positivist scientists would not go much beyond that. They would be happy to find interesting order among the measurement results, and would be satisfied with that. Realist scientists would be happy with that as well, but would not be satisfied. They would look for an underlying reality that brings about the Geiger counter clicks and the dependence of the count rate on distance, intervening material, and time. They would reach an understanding in terms of the sample's consisting of atoms with unstable nuclei, nuclei that spontaneously and unpredictably "decay" and emit particles. Realists would picture the emitted particles entering the counter window and causing the counter to click, one click per particle. They would understand that intervening material absorbs the particles, so that fewer of them would reach the counter window. They would understand that the count rate decreases in time because there remain in the material fewer and fewer undecayed nuclei. They would even suggest new experiments and attempt to predict the results in order to confirm their understanding.

Positivism carried to the extreme leads to *solipsism,* which, in its extreme, is the metaphysical position that nothing is real except the self. In other words, starting with "cogito ergo sum" (I think, therefore I exist), I'm assured only of my own existence; the rest is my imagination (including this book and you readers). It must be admitted, it seems to me, that from a purely logical standpoint, solipsism is the only compelling metaphysical position among them all.

Indeed, the reality to me of everything beyond my own existence is based on arbitrary assumptions that I choose to make.

For example, it is an assumption—perhaps reasonable, maybe not, but nevertheless arbitrary—that my self is somehow intrinsically attached to a body. It is a further assumption that my body is equipped with sense organs and that certain sensations I'm aware of result from the activity of these organs. It is also an assumption that the activity of my sense organs is a result of stimulation by some reality external to my body. And so on and so forth. (Am I just imagining I'm hitting imagined keys that bring these imagined words into imagined existence on an imagined monitor screen? Refer to *Objective and Subjective*, chapter 1.)

Objective Reality

Let's now proceed step by step to see how science leads us to a particular metaphysical position in regard to reality, which, as you can readily predict from the preceding paragraphs, is going to be a version of realism. My arguments and justifications will be based on common sense, on our understanding of nature, and on reasonableness, utility, motivation, and conceptual economy (or parsimony, i.e., thrift in the use of concepts).

To start off, it seems to me that science is urging us to be realists, that we're being encouraged to believe there is an objective reality, an observer-independent underlying reality that is the same reality for electrons, for galaxies, for tadpoles, and for people. One argument here is based on *motivation*. If one doesn't believe there exists an objective reality, there is no motivation to discover and investigate it. Yet the existence of science demonstrates that there are many people who are strongly motivated to discover and investigate objective reality. Since science is so successful in accomplishing its goals, it seems that our motivation to do science is well justified. Thus we're encouraged to believe in an underlying reality. This argument is good for the existence of an underlying reality, but does not address its objectivity.

However, it is most *reasonable* to believe in the objectivity of underlying reality, that it is observer-independent and is the same underlying reality for neutrons, for stars, for carrots, and for humans. Science goes to much trouble and effort to maximize objectivity by confining itself to the reproducible aspects of nature (*Reproducibility*, chapter 2). Then science finds that nature possesses such aspects, and moreover, that there is no lack of raw material for science to process. And finally, almost all this raw material is well comprehended by science and fits together beautifully, meshing into a most elegant conceptual fabric, the whole wonderful world of science. So if it looks, sounds, and smells like objective reality, is it not most *reasonable* and *conceptually economical* to believe that this is just what we have?

Furthermore, it is *useful* to believe there is an objective reality, because, motivated to investigate it, we do investigate it, and, through technological application, we obtain useful results, such as fertilizers, vitamins, telephones, and word processors.

Now that we're convinced to believe in an objective reality, we note that we gain scientific knowledge of it through our physical senses: sight, hearing, smell, taste, touch (pressure), temperature, and others. (Any "knowledge" of reality gained through other, non-physical channels is of no concern to science.) Information from our senses is processed by our nervous system, where it interacts with innate and acquired structures, finally emerging in our consciousness. Thus our knowledge of reality involves interaction with the world external to our bodies, filtering through our senses, neural processing, and conscious awareness of the results of all this.

That seems obvious. The crucial implication is that scientific knowledge of objective reality is *indirect* knowledge. *Direct* knowledge, in contrast, somehow passes directly from the object of the knowledge into our consciousness. We might call it intuition or belief or feeling. But by its character, science confines us to our *physical* senses in our observations of nature.

Here intuition, belief, and feeling are forbidden. "Data" gained by intuiting the reading of a gauge, instead of actually reading it, or by believing a length is such and such, rather than actually measuring it, or by feeling a solution has a certain color, instead of actually seeing it, have no part in science. Thus science dooms itself to merely indirect knowledge of the reality it tries to comprehend.

This puts off some people, who prefer direct knowledge of reality through nonscientific modes of comprehension, in preference to the indirect knowledge of science. Thus they might attempt to know reality by means of meditation, prayer, feeling, intuition, inspiration, awareness, or pure thought, possibly with the help of "awareness-enhancing" or "mind-expanding" drugs. Who but the subject himself or herself can judge whether any knowledge is thus gained at all? And who, including the subject, can judge whether such knowledge has anything whatsoever to do with reality? What can be stated with certainty, though, is that any knowledge thus gained is among the most subjective knowledge imaginable.

Yet scientists, too, indulge in intuition, inspiration, feeling, and pure thought, as we saw in chapter 3, dealing with theories in science. And not only in their devising and judging of theories do scientists act (and must act) with some degree of irrationality, but intuition, inspiration, and taste can also guide the choice of what aspect of nature to investigate, how to go about it, and what experiments to perform, for example. It is to a large extent by their superior intuition and inspiration that the greatest scientists achieve their stature in the science community. But at the level of observational data, no scientist will ever accept "direct knowledge" in lieu of knowledge gained via the physical senses, however indirect the latter admittedly is.

Perceived Reality

What we *do* become aware of via our senses and our neural processing might be termed *perceived reality*. It is perceived reality that is the actual subject of science, although we might hope to gain some understanding of objective reality through an understanding of perceived reality.

We are the result of evolutionary adaptation, according to our *understanding of nature*, and have been managing to survive in our ecological niche for millennia. Thus it is *reasonable* to believe that perceived reality cannot be much out of tune with objective reality, at least with those aspects of reality that strongly and frequently affect our survival as individuals and as a species. When we perceive food at a certain location at a certain time and strive to get hold of it, we succeed in nourishing ourselves often enough to survive. Or, most of us manage to avoid the attacks of tigers (as well as of their spiritual descendants, motor vehicles). Those of us for whom that is not true don't live to pass on their genes.

So, in the absence of any direct scientific knowledge of objective reality, we should not throw up our hands in despair, but rather should take our perceptions and concepts seriously and let them guide us toward an understanding of objective reality. The argument here is one of *motivation* and *utility*: If we have no hope of approaching objective reality, we might as well give up believing in it. It is *reasonable* and *conceptually economical* to believe that our perceptions and concepts even give us a *literal* picture of objective reality, at least as long as this belief is tenable.

Even so, we cannot presume that our perceptions and concepts of reality necessarily give us a literal description of, or even remain faithful guides to, aspects of reality that do not affect us strongly or frequently, such as for submicroscopic phenomena or for astronomical and larger-scale phenomena. The reason is that the argument based on evolutionary adaptation then loses its validity. It does not much matter whether we are adapted to aspects of reality that affect us only weakly or rarely; in any case, those aspects do not significantly affect our survival.

So when dealing with such aspects of reality, it is *reasonable* to let our belief in our ability to describe objective reality literally through our understanding of perceived reality be contingent on its not leading to undue difficulty. We should be prepared to drop our belief in literal description if and whenever it proves to cause more trouble than it is worth.

And it turns out that, when we deal with the quantum aspect of nature, our belief in a literal description of objective reality causes tremendous trouble. Perceived reality at the submicroscopic level of nature, which is well described by quantum theory, is simply unacceptable as a literal description of objective reality. The technicalities of the problem are beyond the scope of this book. (They are presented in other books, some of which are listed in the bibliography at the end of the chapter.) But the cardinal point can be stated without going into technicality. It is that in quantum theory there is too much dependence on the observer to allow perceived reality to be a literal description of objective reality, which is assumed to be independent of observers.

Let's try to clarify that. Quantum theory is formulated in terms of possible happenings and their probabilities of occurring, rather than in strictly deterministic terms of what *will* occur. For example, for some situations quantum theory might tell us that a particle can be in a certain range of positions (rather than specifying its exact location). If we make an observation of its position, we'll find it at some location within the allowed range. But only then, according to quantum theory, are we allowed to think of the particle as actually being located. Prior to the observation the particle must not be thought of as being located at all. It then possesses only potentiality for location. The act of observation realizes the potentiality and endows the particle with the property of position.

It's not as if, prior to the observation, the particle had a location that we didn't know, but which the observation revealed. Quantum theory clearly rules this out and affirms that it is the act of observation itself that bestows the property of position upon the particle, which previously did not possess the property. The intrinsic dependence of the perceived reality of properties, such as position, on the activities of observers demonstrates that the perceived reality of the quantum aspects of nature cannot be a literal description of an observer-independent, objective reality.

That might seem to make little sense, probably even no sense at all. In fact, it might appear to be downright crazy. Yet that's the way things are. The astounding, mind-boggling quantum aspects of nature are counterintuitive. But they are no less part of perceived reality for all that.

The trouble lies with our limited intuition, which developed in the environment of ordinary-size phenomena that affect us strongly and often. At this scale, quantum effects, though valid, are negligible. So we have no intuition for them. Nevertheless, their implications for reality are so crucial that no thinker can afford to ignore them. Thus I repeat my claim (at the beginning of this chapter) that no educational curriculum worthy of its name can do without some minimal exposure to quantum theory (as well as relativity theory). If you're not familiar with the ideas of quantum theory, I strongly urge you to make use of the bibliography entries at the end of the chapter.

Partially Hidden Reality

Thus we know that even if certain aspects of perceived reality can be assumed to give a literal description of objective reality, there are other aspects that do not. We know that science does not give us full comprehension and understanding of objective reality. Whether science ever will do that is a moot issue at present, but the way things are developing does not offer much hope. So the objective reality that science has led us to believe in is partially hidden from us, as a veiled, clouded, fogged reality. Science allows us clear glimpses of parts of it, as well as provocative hints about more of it. Most of objective reality, however, will very likely remain inaccessible to humankind via science.

It seems to be human nature to want what we don't have, and to desire especially what we can't have. And so with objective reality. The simply fascinating questions about it are those we can't answer at present, whereas the profound ones will most probably never be answered by science. Questions such as: What can be deduced from quantum theory about objective reality? Are space and time, as we perceive them, fundamental aspects of objective reality? (There appear to be indications to the negative.) And if not, how do they link to it? To what extent is our perception of nature in terms of localized objects valid for objective reality? (Not very, it seems.) What relation does mind bear to objective reality? And the profoundest questions of all are most likely far beyond human mental ability even to formulate.

Since science does not, and most probably cannot, give us anything near full comprehension and understanding of objective reality, some claim that there seems to be no reason why other, nonscientific modes of comprehension and understanding should not afford us hints and clues to this reality. It's claimed that as long as objective reality, which science guides us to believe in, itself lies beyond the domain of science, we should not shut ourselves off from the possibility that we might possess other channels to it. Indeed, it is claimed, why shouldn't intuition, belief, feeling, art, music, and poetry be allowed to contribute whatever insight they might offer? Such modes of comprehension, it is claimed, as irrational and subjective as they are, and perhaps even *because* of their irrationality and subjectivity, might complement, and thus strengthen, the contribution of science to our quest for objective reality.

Might or might not, I reply. I try to remain open-minded, but am skeptical. After all, objective is objective, and it's hard to see how the subjective and irrational can contribute to an understanding of the objective.

Let's summarize. Using arguments and justifications based on common sense, on our understanding of nature, and on reasonableness, utility, conceptual economy, and motivation, we have seen how science leads us to a belief in an objective reality. Our scientific knowledge of this reality is indirect. But science should nevertheless give us a literal description of objective reality, at least for those aspects of it that strongly and frequently affect our survival as individuals and as a species. Yet quantum theory, the best theory we presently possess, cannot be a literal description of this reality. Thus objective reality is partially, very likely mostly, hidden from us.

Transcendent Reality

In earlier chapters we made much of the concept of nature. How does nature accord with reality? Recall that nature is, for our purpose, the material universe with which we can, or can conceivably, interact. That's a close match with perceived reality. Science, we recall, is our attempt to understand objectively the reproducible and predictable aspects of nature, and thus of perceived reality. (In *Perceived Reality*, this chapter, I pointed out that science is concerned with perceived reality.) Both nature and perceived reality are phenomena of objective reality—which transcends and subsumes nature.

Earlier in the book (in *Transcendence and Nontranscendence*, chapter 4), I declared that I hold a nontranscendent worldview, one that makes do without any nature-transcending reality. In the present chapter we saw how science guides us to a worldview of partially hidden objective reality, and as might be guessed, I subscribe to this worldview. Since objective reality transcends nature, have I then run into a contradiction? Yes, I have. At the stage of our investigation that we discussed in *Transcendence and Nontranscendence*, chapter 4, I wanted to hold as simple a worldview as seemed warranted by science, thus a nontranscendent one. I assumed, as do many scientists, that we would thereby find the objectivity we strive for through science. Transcendence was felt to smack too much of subjectivity.

But it turns out, as we just saw, that science, in its study of nature, cannot fulfill all of our demands for objectivity. The quantum aspect of nature involves observers too much for that. So any objective reality must be "farther" from us than nature, "more distant" from us than perceived reality. It must transcend them. Thus I'm led to the belief in a partially hidden objective reality, a reality transcending nature, a reality most likely surpassing human understanding. But nonetheless an objective reality! An observer-independent underlying reality that is the same reality for protons, for planets, for peanuts, and for people.

Is there room in it for God? Deities? Omniscience? Omnipotence? Supernatural powers? Grand designs? Perhaps for some of that, since once Pandora's box of transcendence is opened, it's hard to say what exactly is allowed and what is not. After all, we are talking about a metaphysical position (*Science and Metaphysics*, chapter 4), albeit a "scientific" one, so we are free to assume whatever we want, as long as consistency with science is maintained. And, let me emphasize, as long as objectivity is maintained; as long as one's assumptions do not contradict observer independence, or the idea that it is the same underlying reality for helium, for helicopters, for hyacinths, and for all humans. If objectivity were to be abandoned, there would be no point in the whole metaphysical construct we have developed in this chapter. And self-consistency? As one likes, since we don't really expect to fully understand objective reality anyhow.

𝒜

Using arguments and justifications based on common sense, on our understanding of nature, and on reasonableness, utility, conceptual economy, and motivation, we're led by science to a belief in an objective reality. Our scientific knowledge of this reality, however, is gathered indirectly. But science should nevertheless give us a literal description of objective reality, at least for those aspects of it that strongly and frequently affect our survival as individuals and as a species. Yet quantum theory cannot be a literal description of objective reality. Thus objective reality is partially, very likely mostly, hidden from us. Since this reality transcends nature, we are thus led to a transcendent worldview.

Bibliography

The principal references for this chapter are:

B. d'Espagnat, *Reality and the Physicist: Knowledge, Duration and the Quantum World* (Cambridge University Press, Cambridge, 1989).

———, *In Search of Reality* (Springer-Verlag, New York, 1983).

Neither of these is easy reading, the 1989 book harder than the 1983 one. This chapter is in essence an adaptation, with twists of my own, of d'Espagnat's ideas concerning what he aptly calls "veiled objective reality." That will become obvious upon reading his book(s). For more books dealing in various ways with reality, see

R. K. Adair, *The Great Design: Particles, Fields, and Creation* (Oxford University Press, Oxford, 1987).

M. Eigen and R. Winkler, *Laws of the Game: How the Principles of Nature Govern Chance* (Knopf, New York, 1981, and Princeton University Press, Princeton, N.J., 1993).

H. Fritzsch, *The Creation of Matter: The Universe from Beginning to End* (Basic Books, New York, 1984).

E. Harrison, *Masks of the Universe: Changing Ideas on the Nature of the Cosmos*, 2nd ed. (Cambridge University Press, Cambridge, 2003).

K. Lorenz, *Behind the Mirror: A Search for a Natural History of Human Knowledge* (Harcourt Brace Jovanovich, New York, 1977).

R. Morris, *The Nature of Reality* (McGraw-Hill, New York, 1987).

———, *Dismantling the Universe: The Nature of Scientific Discovery* (Simon and Schuster, New York, 1983).

D. Park, *The How and the Why* (Princeton University Press, Princeton, N.J., 1988).

A. Rae, *Quantum Physics: Illusion or Reality* (Cambridge University Press, Cambridge, 1986).

J. S. Trefil, *Reading the Mind of God: In Search of the Principle of Universality* (Charles Scribner's Sons, New York, 1989).

A. Zee, *Fearful Symmetry: The Search for Beauty in Modern Physics* (Macmillan, New York, 1986, and Princeton University Press, Princeton, N.J., 2007).

Some suggested books on metaphysics are

E. Conee and T. Sider, *Riddles of Existence: A Guided Tour of Metaphysics* (Oxford University Press, New York, 2007).

A. Gianelli, K. Kennedy, and G. Statile, eds., *The Journey of Metaphysics*, 2nd ed. (Pearson, Boston, 2006).

M. Lange, *An Introduction to the Philosophy of Physics: Locality, Fields, Energy, and Mass.* (Wiley-Blackwell, Hoboken, N.J., 2002).

P. van Inwagen, *Metaphysics*, 3rd ed. (Westview, Boulder, Colo., 2008).

For adaptive biological evolution, see

J. Monod, *Chance and Necessity: An Essay on the Natural Philosophy of Modern Biology* (Knopf, New York, 1971).

Also Lorenz, and Eigen and Winkler, in the lists above.

For the observer dependence of quantum reality, see

J. D. Barrow, *The World within the World* (Oxford University Press, Oxford, 1988).

P. C. W. Davies, *God and the New Physics* (Simon and Schuster, New York, 1983).

————, *Other Worlds: A Portrait of Nature in Rebellion; Space, Superspace and the Quantum Universe* (Simon and Schuster, New York, 1980).

P. C. W. Davies and J. R. Brown, eds., *The Ghost in the Atom: A Discussion of the Mysteries of Quantum Physics* (Cambridge University Press, Cambridge, 1986).

H. R. Pagels, *The Cosmic Code: Quantum Physics as the Language of Nature* (Simon and Schuster, New York, 1982).

J. C. Polkinghorne, *The Quantum World* (Princeton University Press, Princeton, N.J., 1984).

F. Rohrlich, *From Paradox to Reality: Our New Concepts of the Physical World* (Cambridge University Press, Cambridge, 1987).

See also Adair, and Rae, in the lists above.

Coda

SO WHERE ARE WE? What have we accomplished? What have we learned from our discussions in this book?

We started off with an analysis of objectivity and subjectivity. We based it on the definition of "objective" as existing as part of reality, independent of thought or of an observer, and the definition of "subjective" as existing in the mind, belonging to the thinking subject rather than to the object of thought. We considered external reality, the objective outer world, and saw how we gain knowledge of it through our physical senses and by means of social interaction. Thus this knowledge is actually intersubjective, but we referred to it as objective for the sake of our discussion. Then we looked at our individual subjective inner worlds, the worlds of our fantasies, emotions, feelings, perceptions, beliefs, and so on. Subjective inner worlds are incommensurate with one another.

We discussed three kinds of truth: objective, subjective, and logical. Objective truth is consistency with reality, passing the reality check, and is in the public domain. Subjective truth, however, is strictly a private affair; its domain of validity is confined to the subjective inner world of an individual. Thus, for example, ostensibly conflicting beliefs, if they do not deal with the real world (if they did, they could be tested and compared), are not in contradiction. Logical truth is the truth of logical deductions, including mathematical theorems, within the self-consistent logical structures in which they are derived. A logical truth may also be an objective truth, although it does not need to be.

We devoted some discussion to dealing with the subjective. Dealing with the objective real world, however, is best done through science. Objectivity is a goal of science, as indicated by the definition of science as our attempt to understand objectively the reproducible and predictable aspects of nature.

Then, based on this definition, we considered what science is about and how it operates. Taking nature as the material universe with which we can, or can conceivably, interact, we saw that in doing science, we first search for order among the reproducible phenomena of nature. We then attempt to formulate laws that describe the collected data and predict new results. After finding such laws, we try to explain them by means of theories.

In this connection, we studied what reproducibility and predictability are, and their relation to order and law, where laws are descriptions of and abstractions from order. We considered Kepler's laws of planetary motion as an archetypal example of laws of nature. An objective explanation of a law of nature is a theory.

We looked into what makes a theory acceptable, and examined the properties of theories that scientists consider to be advantageous. For a theory to be acceptable, it must explain what it is designed to explain (the theory must be objectively true) and what is doing the explaining must logically imply what is being explained (the theory must be logically true). In addition, the following are advantageous: What is explaining should be as much an aspect of nature as what is being explained (naturality). The former should also be more general (generality), more fundamental (fundamentality), more unifying (unification), and simpler (simplicity) than the latter and should be perceived as causing the latter (causation). Theories should be falsifiable (falsifiability), in the sense that they offer predictions that allow them to be tested. Beautiful theories are preferred.

We considered Newton's laws as an archetypal example of a theory, in this case a theory to explain Kepler's laws.

After dealing with science, we briefly examined what metaphysics, in the sense of the philosophical framework in which science operates, is about. We compared science and metaphysics and their respective domains, and noted that, in contrast to science, metaphysics lacks criteria for objective truth. Thus worldviews, which are conceptual frameworks and lie within the domain of metaphysics, are highly subjective and personal. Then we considered two particular types of worldview and compared them: transcendent worldviews, which involve the existence of a reality beyond nature, and nontranscendent worldviews, which make do with nature as all there is.

We examined the character of the universe *as a whole* and recognized that, as far as science is concerned, the universe is a unique phenomenon. As such, the universe is intrinsically irreproducible and thus lies outside the framework of science. As a result, order, law, predictability, and theory are irrelevant to the universe *as a whole*; it is orderless, lawless, unpredictable, and unexplainable through science.

It follows that cosmological schemes, such as the inflationary big-bang scenarios, as useful and valuable as they are, cannot be and are not theories: they are attempts to *describe*, not scientifically *explain*, the working of the universe. Hence cosmology, to the extent it attempts to deal with the universe as a whole, is basically metaphysics.

Since the universe as a whole is orderless, lawless, unpredictable, and unexplainable, how is it that the universe possesses aspects and phenomena that *are* orderly, lawful, predictable, and explainable? How is it that within our lawless universe there are laws of nature? In other words, how can we be doing science? This question was the subject of our subsequent discussion, which led us through a number of diverse considerations.

We compared the metaphysical positions of realism and idealism. Realism holds that the order and laws of nature we find are really "out there," objective and independent of observers, whereas idealism claims that the order and laws of nature are wholly subjective, in the mind of the observer. Then we considered the hybrid position that order is an objective property of nature, but laws are mental constructs.

We examined and compared the worldviews of holism and reductionism. The former states that nature can be understood only in its wholeness or not at all. The latter claims that nature is understandable as the sum of its parts and should be studied by analysis and synthesis. We saw that although nature is an integrated whole, reductionism nonetheless succeeds to a great extent, since science, which generally operates reductionistically, is successful in achieving a good understanding of many aspects and phenomena of nature. Yet nature does possess nonseparable aspects as well.

We examined three ways in which science reduces nature to its parts, as well as the limitations of the three ways, to wit:

1. Observer and observed, which works well down to the limit set by nature's intrinsic quantum nonseparability, that is, it works for phenomena that are not too small.
2. Quasi-isolated system and surroundings, where a quasi-isolated system is a system that is isolated from its surroundings to the best of our ability and understanding. Such isolation is necessarily imperfect, due to inertia and the uncontrollable quantum correlations involved in nature's nonseparability. Nature's order is manifested in quasi-isolated systems, and it is for them that laws of nature are found.
3. Initial state and law of evolution. This reduction allows lawfulness to be found for the time development of quasi isolated systems, but it is not applicable to the universe as a whole.

We considered the Mach principle, that the origin of inertia lies with all the matter in the universe, and also its generalization to the extended Mach principle, that the origin of the laws of nature for quasi-isolated systems lies with the universe as a whole. Thus the orderless, lawless, unpredictable, and unexplainable universe not only possesses, but can actually be thought of as engendering, orderly, lawful, predictable, and explainable behavior in aspects and parts of itself. Still, nature's orderlessness reveals itself on various scales. Science can offer no conventional explanation for the existence of laws of nature within the lawless universe.

We turned to an examination of science as a human endeavor: (1) science is a by-product or *our* existence; (2) it is *our* conception of nature that we are attempting to explain; and (3) a valid explanation is one that satisfies *us*. That led us to the anthropic principle, which states that the existence of *Homo sapiens* may, within the framework of science, serve as an explanation for phenomena and aspects of nature, and moreover that such explanations are the most fundamental. Fundamentality follows from the fact that our existence is the most fundamental aspect of nature *for us*.

But anthropic explanations should be used only as a last resort, and only for aspects of nature that are apparently so fundamental that no conventional explanation is available. One reason this is so is because they suffer from the subjective difficulty that our existence seems neither more general, more unifying, nor more simple than whatever it is explaining, and neither is it perceived as causing whatever it is explaining. Anthropic explanations suffer also from the objective difficulty of the invariant-context problem, when they are used by comparing the actual state of the universe with hypothetical alternative situations and studying how the alternatives would affect our existence. Such a use involves the hypothetical variation of some aspect of nature, while keeping all other aspects unchanged. But—and this is the invariant-context problem—we can never be sure the variation cannot be compensated for by concomitant change in other aspects of nature.

Since science offers no conventional explanation for the existence of order within the orderless universe, we turned in desperation to the anthropic principle, which, as we saw, allows a rather weak anthropic explanation. But we found a clean, unqualified anthropic explanation for space and time, for which no conventional explanation seems to be available.

Our final discussion in this book concerned the nature of reality, in particular what science tells us about it. Using arguments and justifications based on common sense, on our understanding of nature, and on reasonableness, utility, conceptual economy, and motivation, we're led by science to the belief in an objective reality. We saw, however, that our scientific knowledge of this reality is indirect. But science should nevertheless give us a literal description of objective reality, at least for those aspects of it that strongly and frequently affect our survival as individuals and as a species. Yet quantum theory, which is the best theory we have, cannot be a literal description of this reality. Thus objective reality is at least partially hidden from us. Since this reality transcends nature, we're thus led to a transcendent worldview. There's more to objective reality than meets the eye, it seems.

GLOSSARY

Terms in **bold type** within entries are to be found as their own glossary entries.

anthropic principle The existence of Homo sapiens may, within the framework of **science**, serve as an explanation for phenomena and aspects of **nature**, and, moreover, such explanations are the most fundamental.

beauty As referred to a **theory**, a beautiful theory is one that arouses the aesthetic feeling of beauty in the scientist considering it. The feeling of beauty is found to be engendered by the properties of simplicity, unification, and generality, when possessed by the theory. Scientists, in their belief that **nature** should be understandable in terms of beautiful theories, tend to prefer more beautiful theories to less beautiful ones, even at the expense of **objective** advantages. Amazingly, successful theories do tend to be beautiful.

big bang The cosmic explosion, or primeval fireball, by which, according to certain cosmological schemes [**scheme**], the universe came into existence and has been expanding ever since.

big crunch The cosmic implosion, or utter collapse, by which, according to certain cosmological schemes [**scheme**], the universe will go out of existence after passing through a contraction era following the present expansion era.

biological evolution As explained by Darwin's theory and its subsequent modifications, the evolved fitness of organisms, especially of humans, forms a major argument in the discussions presented in this book.

161

conservatism   In science: Hold on to what you have, stick to the tried and well-confirmed, for as long as is reasonably possible; make change only when the need for change becomes overwhelming; and then make only the minimal change needed to achieve the desired end. Specifically, with regard to the laws of nature [law]: as long as there is no compelling reason to the contrary, assume that the laws of nature we find here and now are, were, and will be valid everywhere and forever.

cosmological scheme.   See scheme.

cosmology   The study of the working of the cosmos, the universe as a whole, at present, in the past, and in the future. In its dealings with the connections and interrelations among the aspects and phenomena of the universe, cosmology can be considered a branch of science. But in its holistic [holism] mode, when it attempts to comprehend the universe as a whole, it is a branch of metaphysics.

dimension   For our discussion, the possibility of assigning a measure to something, as length, width, and height are dimensions of a box. (See space and time.)

elementary particle   Any of many kinds of subatomic and subnuclear particles, including the electron, positron, proton, neutron, neutrino, pion, etc. The name is historic, as the actual elementarity of many of the so-called elementary particles is obviated by present-day understanding, which views them as being composed of more-elementary constituents.

evolution, biological.   See biological evolution.

evolution, law of.   See law of evolution.

evolution (of nature)   The process of nature's [nature] change in time.

extended Mach principle   The origin of the laws of nature [law] for quasi-isolated systems [quasi-isolated system] lies with the universe as a whole. (See also March principle.)

falsifiability   Testability. The property of a theory that it can be tested against as yet unknown natural [nature] phenomena to determine whether it is true or false. In order to be falsifiable, a theory must predict [predictability] something in addition to what it originally explained.

holism The metaphysical [**metaphysics**] position that **nature** can be understood only in its wholeness, including human beings, or not at all. (Compare with **reductionism**.)

idealism The metaphysical [**metaphysics**] position that the laws of nature [**law**] are not inherent to the external world, i.e., to **objective reality**, but are mental constructs, artifacts of the way our minds interpret and organize our sensory impressions, thus of the way we perceive the world. (Compare with **realism**.)

inertia The property of bodies according to which a body's behavior is governed by Newton's first universal law of motion [**Newton's laws**]: In the absence of forces acting on them or when such forces cancel each other, bodies remain at rest or continue to move uniformly in a straight line. (See **Mach principle**.)

inflation According to certain cosmological schemes [**scheme**], an era of unimaginably rapid expansion of the universe (called the inflationary era) starting soon after the **big bang**.

initial state Any state of a material system, when considered as a precursor state from which the subsequent **evolution** of the system follows. (See **law of evolution**.)

isolated system The idealization of a material system that has absolutely no interaction with the rest of the universe. An isolated system would not be part of **nature**, since we could not interact with it and hence could not observe it. (Compare with **quasi-isolated system**.)

Kepler's laws Johannes Kepler's three laws [**law**] of planetary motion:

1. The path each planet traverses in **space**, its orbit, lies wholly in a fixed plane and has the form of an ellipse, of which the Sun is located at a focus.
2. As each planet moves along its elliptical orbit, the (imaginary) line connecting it with the Sun sweeps out equal areas during equal time intervals.
3. The ratio of the squares of the orbital periods of any two planets equals the ratio of the cubes of their respective orbital major axes.

The laws express an **order** among the astronomical data
concerning the planets and offer a description and a unification of
the data. They predicted the relevant properties of the motions of
the planets discovered after Kepler's time. They are also valid for
any system of astronomical bodies revolving around a massive
central body, such as the moons of the planet Jupiter.

law (of nature) An expression of **order**, thus of simplicity, in **nature**.
A compact condensation of all relevant existing data, as well as of
any amount of potential data, a law is a unifying, descriptive
device for its relevant class of natural phenomena. Laws enable us
to predict the results of new experiments (see **predictability**).

law of evolution Any **law** of nature that, given any **initial state** of a
quasi-isolated system, yields the state that evolves [**evolution**] from
it at any subsequent **time.**

law of nature. See **law (of nature).**

laws, Kepler's. See **Kepler's laws.**

laws, Newton's. See **Newton's laws.**

Mach principle The origin of **inertia** lies with all the matter of the
universe. (See also **extended March principle.**)

Mach principle, extended. See **extended Mach principle.**

many worlds The many-worlds interpretation of **quantum theory**
holds that at every instant the universe "branches" into realizations
of all the quantum possibilities of that instant, which continue to
"coexist side-by-side," each branch universe branching further at
the next instant, and so on.

metaphysics A branch of philosophy dealing with being and reality.
Metaphysics, as used in this book, is the philosophic framework in
which **science** operates. In this sense, metaphysics is concerned
with what lies around, below, above, before, and beyond science. A
metaphysical position is part of one's **worldview.**

metatime A metaphysical [**metaphysics**] **time** "higher" than ordinary
time, in which universes can be born and die.

nature The material universe with which we can, or can conceivably,
interact; i.e., everything of purely material character that we can,
or can conceivably, observe and measure. By "conceivably" we

mean that it is not precluded by any principle known to us, and is considered attainable through further technological research and development. We exclude from nature such concepts as mind, idea, feeling, consciousness, etc.

Newton's laws   Isaac Newton's three universal laws [**law**] of motion:

(1) In the absence of forces acting on it or when such forces cancel each other, a body will remain at rest or continue to move uniformly in a straight line.

(2) A force acting on a body will cause the body to undergo acceleration whose direction is that of the force, and whose magnitude is proportional to that of the force divided by the body's mass.

(3) For every force acting on it, a body will react upon the force's source with a force of opposite direction and equal magnitude along the same line of action.

Newton's law of universal gravitation: Every pair of bodies undergoes mutual attraction, such that the force acting on each body is proportional to the product of the bodies' masses and inversely proportional to the square of their separation.

These four laws form Newton's **theory** to explain **Kepler's laws** of planetary motion, as well as a vast realm of other mechanical phenomena.

nontranscendent worldview.   See **worldview, nontranscendent.**

objective   Existing as part of reality, independent of thought or of an observer.

objective reality   The **observer**-independent reality underlying the phenomena of **nature. Science** shows that **objective** reality must be partially, very likely mostly, hidden from us. (Compare with **perceived reality.**)

observed   The rest of **nature,** as distinct from us (*Homo sapiens*), as **observer,** in the observer-observed separation [**separability**] of nature, according to the reductionist [**reductionism**] approach.

observer   We (*Homo sapiens*), as distinct from the rest of **nature,** as **observed,** in the observer-observed separation [**separability**] of nature, according to the reductionist [**reductionism**] approach.

order   The opposite of randomness, of haphazardness. The existence of relations among natural [**nature**] phenomena. Order is a simplicity of nature, whereby phenomena that might otherwise have had nothing to do with each other (and would have presented a more complex situation) are interrelated and interdependent (yielding a situation that is less complex).

particle, elementary.   See **elementary particle**.

perceived reality   That which we become aware of via our physical senses. Perceived reality is the actual subject of **science**. (Compare with **objective reality**.)

phenomenon, unique.   See **unique phenomenon**.

positivism   The metaphysical [**metaphysics**] position that only our sense data, derived from measurements and observations, are fundamental. (Compare with **realism**.)

predictability   The characteristic that among the natural [**nature**] phenomena investigated, **order** can be found, from which laws [**law**] can be formulated that predict the results of new experiments. Predictability makes **science** a means both to understand and to exploit nature. We don't claim that nature is predictable in all its aspects, but any unpredictable aspects it might possess lie outside the domain of science.

quantum theory   A very formal and mathematical **theory** concerned with the fundamental behavior of all material systems in principle, but usually and most usefully applied to molecular, atomic, and subatomic systems. Quantum theory is formulated in terms of possible happenings and their probabilities of actually occurring (rather than in strictly deterministic terms of what *will* occur). Individual submicroscopic events, according to this theory, are inherently unpredictable [**predictability**]; it is only their probability that can be predicted. Quantum theory implies, among other things, that **nature** is fundamentally nonseparable [**separability**], so that uncontrollable correlations can exist among ostensibly separated systems. Quantum theory also implies that **perceived reality** is **observer**-dependent, thus that perceived reality cannot be a literal description of **objective reality**.

quasi-isolated system   Any material system that is as nearly isolated as possible from the rest of the universe, i.e., whose interaction with the rest of the universe is reduced to the minimum possible. (Compare with **isolated system**; see **surroundings**.) It is for quasi-isolated systems that laws of nature [**law**] are found.

realism   In general, the metaphysical [**metaphysics**] position that there exists an underlying **observer**-independent, **objective reality**; that **nature** would manage just as well if we weren't around. Also specifically, the metaphysical position that the laws of nature [**law**] reside in this reality. (Compare with **positivism** and with **idealism**.)

reality, objective.   See **objective reality**.

reality, perceived.   See **perceived reality**.

reductionism   The metaphysical [**metaphysics**] position that **nature** can be understood as the sum of its parts and thus should be studied by analysis and synthesis. (Compare with **holism**; see **separability**.)

reproducibility   The possibility of repeating experiments by the same and other investigators, thus yielding data of **objective**, lasting value about the phenomena of **nature**. Reproducibility makes **science** a common human endeavor, and, as nearly as possible, an **objective** endeavor of lasting validity. Nature is not claimed to be reproducible in all its aspects, but any irreproducible aspects it might possess lie outside the domain of science.

scheme   Especially as cosmological [**cosmology**] scheme, an attempt to *describe* the working of the cosmos, the universe as a whole. Cosmological schemes are not theories [**theory**], do not *explain* the working of the cosmos, since the universe as a whole, being a **unique phenomenon** and thus irreproducible [**reproducibility**], lies outside the domain of **science**.

science   Our attempt to objectively understand, i.e., to be able to objectively explain, the reproducible [**reproducibility**] and predictable [**predictability**] aspects of **nature**. Science is a human endeavor, since *Homo sapiens* plays an essential role in it, in that: (1) science is a by-product of *our* existence; (2) it is *our* conception of nature that we're attempting to explain; and (3) a valid explanation in science is one that satisfies *us*.

separability The amenability of **nature** to our attaining an understanding of it through analysis and synthesis. **Reductionism** holds that separability is valid for nature.

space The **dimension** of being, of existence; the possibility of assigning a measure (consisting of three numbers) to being.

space-time **Space** and **time** together, considered as a single concept. Einstein's theories [**theory**] of relativity are formulated in terms of space-time.

subjective Existing in the mind; belonging to the thinking subject rather than to the object of thought.

surroundings (of quasi-isolated system) The rest of the universe, as distinct from any **quasi-isolated system**.

system, isolated. See **isolated system**.

system, quasi-isolated. See **quasi-isolated system**.

theory A scientific [**science**] explanation of a law of nature [**law**]. A theory gives reasons for the **law** it explains. The acceptability of a theory depends on its giving the feeling that something is indeed being explained. This feeling is found to be fostered by the existence of a number of properties of a theory, the essential one being that whatever is doing the explaining must logically imply that which is being explained. Additional properties enhancing a theory's acceptability are: what is doing the explaining must be as much an aspect of nature as what is being explained, and the former must be more general, more fundamental, more unifying, and simpler than the latter, and be perceived as causing the latter. Beautiful [**beauty**] theories are preferred. A theory should be falsifiable [**falsifiability**].

theory, quantum. See **quantum theory**.

Theory of Everything (TOE) The general name for any putative **theory** explaining the universe as a whole. A TOE, however, were one to be devised, could not be a theory, could not be an *explanation*, since the universe as a whole lies outside the domain of **science**. It would be a **scheme**, a *description* of the universe as a whole.

time The **dimension** of becoming, of change; the possibility of assigning a measure (consisting of a single number) to becoming.

transcendent worldview.   See **worldview, transcendent.**

unique phenomenon   A natural [**nature**] phenomenon that is *essentially* different from all other natural phenomena. Thus a unique phenomenon is intrinsically irreproducible [**reproducibility**]. The universe as a whole is an example par excellence of a unique phenomenon.

worldview   One's attitude toward and interpretation of reality; the conceptual framework by which one organizes one's perceptions.

worldview, nontranscendent   Any **worldview** that excludes a reality beyond, or transcending, **nature** and makes do with nature as all there is. Nontranscendent worldviews do not necessarily deny the existence of mind, consciousness, thought, emotion, feeling, etc. (Compare with **worldview, transcendent.**)

worldview, transcendent   Any **worldview** involving the existence of a reality beyond, or transcending, **nature**. Nature, with which **science** is concerned, is viewed as being embedded in, being part of, transcendent reality. A religion is an example of a transcendent worldview. (Compare with **worldview, nontranscendent.**)

COMBINED BIBLIOGRAPHY

What follows is a compilation of the books listed in all of the chapter bibliographies, arranged alphabetically by author. The numbers concluding the entries here are the numbers of the chapters in whose chapter bibliographies a book is listed.

R. K. Adair, *The Great Design: Particles, Fields, and Creation* (Oxford University Press, Oxford, 1987). 2, 5–8

P. W. Atkins, *The Creation* (Freeman, San Francisco, 1981). 5

J. D. Barrow, *The World within the World* (Oxford University Press, Oxford, 1988). 2, 4–8

J. D. Barrow and J. Silk, *The Left Hand of Creation: The Origin and Evolution of the Expanding Universe* (Heinemann, London, 1983). 5–7

J. D. Barrow and F. J. Tipler, *The Anthropic Cosmological Principle* (Oxford University Press, Oxford, 1986). 7

H. Broch, *Exposed! Ouija, Firewalking, and Other Gibberish* (Johns Hopkins University Press, Baltimore, 2009). 2, 5–7

G. Charpak and H. Broch, *Debunked! ESP, Telekinesis, and Other Pseudoscience* (Johns Hopkins University Press, Baltimore, 2004). 2, 5–7

F. Close, *The Cosmic Onion: Quarks and the Nature of the Universe* (American Institute of Physics, New York, 1983). 5, 6

E. Conee and T. Sider, *Riddles of Existence: A Guided Tour of Metaphysics* (Oxford University Press, New York, 2007). 4, 8

A. H. Cromer, *Connected Knowledge: Science, Philosophy, and Education* (Oxford University Press, Oxford, 1997). 1, 4

———, *Uncommon Sense: The Heretical Nature of Science* (Oxford University Press, Oxford, 1993). 1

G. Darvas, *Symmetry* (Birkhäuser, Basel, 2007). 2, 3

P. C. W. Davies, *The Forces of Nature*, 2nd ed. (Cambridge University Press, Cambridge, 1986). 6

———, *God and the New Physics* (Simon and Schuster, New York, 1983).
4–6, 8

———, *The Accidental Universe* (Cambridge University Press, Cambridge, 1982). 6, 7

———, *The Edge of Infinity* (Simon and Schuster, New York, 1981). 5

———, *Other Worlds: A Portrait of Nature in Rebellion; Space, Superspace and the Quantum Universe* (Simon and Schuster, New York, 1980). 5–8

———, *The Runaway Universe* (Harper and Row, New York, 1978). 5

———, *Space and Time in the Modern Universe* (Cambridge University Press, Cambridge, 1977). 5, 7

P. C. W. Davies and J. R. Brown, eds., *Superstrings: A Theory of Everything?* (Cambridge University Press, Cambridge, 1988). 5

———, eds., *The Ghost in the Atom: A Discussion of the Mysteries of Quantum Physics* (Cambridge University Press, Cambridge, 1986). 5, 6, 8

B. d'Espagnat, *Reality and the Physicist: Knowledge, Duration and the Quantum World* (Cambridge University Press, Cambridge, 1989). 8

———, *In Search of Reality* (Springer-Verlag, New York, 1983). 6–8

M. Eigen and R. Winkler, *Laws of the Game: How the Principles of Nature Govern Chance* (Knopf, New York, 1981, and Princeton University Press, Princeton, N.J., 1993). 4, 6–8

T. Ferris, *The Whole Shebang: A State-of-the-Universe(s) Report* (Simon and Schuster, New York, 1997). 5

R. P. Feynman, *The Character of Physical Law* (MIT Press, Cambridge, Mass., 1965). 2, 5, 6

H. Fritzsch, *The Creation of Matter: The Universe from Beginning to End* (Basic Books, New York, 1984). 2, 4–8

A. Gianelli, K. Kennedy, and G. Statile, eds., *The Journey of Metaphysics*, 2nd ed. (Pearson, Boston, 2006). 4, 8

M. Gleiser, *The Dancing Universe: From Creation Myths to the Big Bang* (Dutton (Penguin), New York, 1997). 5

S. J. Gould, *Science and Religion in the Fullness of Life* (Random House, New York, 1998). 1

B. Greene, *The Fabric of the Cosmos: Space, Time, and the Texture of Reality* (Vintage, New York, 2004). 5

———, *The Elegant Universe: Superstrings, Hidden Dimensions, and the Quest for the Ultimate Theory* (Vintage, New York, 2003). 5

J. Gribbin, *The Search for Superstrings, Symmetry, and the Theory of Everything* (Back Bay, Newport Beach, Calif., 2000). 5

J. Gribbin and M. Rees, *The Stuff of the Universe: Dark Matter, Mankind and the Coincidences of Cosmology* (Heinemann, London, 1989). 5, 7

P. R. Gross, N. Levitt, and M. W. Lewis, eds., *The Flight from Science and Reason* (New York Academy of Sciences, New York, 1996). 1

A. Guth, *The Inflationary Universe: The Quest for a New Theory of Cosmic Origins* (Addison-Wesley, Reading, Mass., 1997). 5

E. Harrison, *Masks of the Universe: Changing Ideas on the Nature of the Cosmos*, 2nd ed. (Cambridge University Press, Cambridge, 2003). 6–8

S. W. Hawking, *A Brief History of Time: From the Big Bang to Black Holes* (Bantam, London, 1988). 5–7

T. Hey and P. Walters, *The Quantum Universe* (Cambridge University Press, Cambridge, 1987). 5, 6

P. Kurtz, ed., *Science and Religion: Are They Compatible?* (Prometheus, Amherst, N.Y., 2003). 1

M. Lange, *An Introduction to the Philosophy of Physics: Locality, Fields, Energy, and Mass.* (Wiley-Blackwell, Hoboken, N.J., 2002). 4, 8

K. Lorenz, *Behind the Mirror: A Search for a Natural History of Human Knowledge* (Harcourt Brace Jovanovich, New York, 1977). 6–8

J. Monod, *Chance and Necessity: An Essay on the Natural Philosophy of Modern Biology* (Knopf, New York, 1971). 4, 6–8

R. Morris, *The Nature of Reality* (McGraw-Hill, New York, 1987). 6–8

———, *Dismantling the Universe: The Nature of Scientific Discovery* (Simon and Schuster, New York, 1983). 2, 6–8

———, *The Fate of the Universe* (Playboy Press, New York, 1982). 5

———, *The End of the World* (Anchor Press, Garden City, N.Y., 1980). 5

R. G. Newton, *The Truth of Science: Physical Theories and Reality* (Harvard University Press, Cambridge, Mass., 1997). 1, 2

———, *What Makes Nature Tick?* (Harvard University Press, Cambridge, Mass., 1993). 2, 3

H. R. Pagels, *Perfect Symmetry: The Search for the Beginning of Time* (Simon and Schuster, New York, 1985, and Bantam, Toronto, 1986). 5

———, *The Cosmic Code: Quantum Physics as the Language of Nature* (Simon and Schuster, New York, 1982). 2, 5, 6, 8

D. Park, *The How and the Why* (Princeton University Press, Princeton, N.J., 1988). 2, 3, 5–8

J. C. Polkinghorne, *Belief in God in an Age of Science* (Yale University Press, New Haven, Conn., 1998). 1, 4

———, *One World: The Interaction of Science and Theology* (Princeton University Press, Princeton, N.J., 1986). 4

———, *The Quantum World* (Princeton University Press, Princeton, N.J., 1984). 5, 6, 8

I. Prigogine and I. Stengers, *Order out of Chaos: Man's New Dialogue with Nature* (Bantam Books, New York, 1984). 2, 4, 6, 7

A. Rae, *Quantum Physics: Illusion or Reality* (Cambridge University Press, Cambridge, 1986). 5, 6, 8

F. Rohrlich, *From Paradox to Reality: Our New Concepts of the Physical World* (Cambridge University Press, Cambridge, 1987). 3, 5–8

J. Rosen, *Symmetry Discovered: Concepts and Applications in Nature and Science* (Cambridge University Press, Cambridge, 1975; reprinted with additions by Dover Publications, Mineola, N.Y., 1998). 2, 3

C. Sagan, *The Demon-Haunted World: Science as a Candle in the Dark* (Ballantine Books, New York, 1996). 1

M. Shermer, *Why People Believe Weird Things: Pseudoscience, Superstition, and Other Confusions of Our Time* (Freeman, San Francisco, 1997). 1, 2, 4–7

J. Silk, *The Big Bang*, revised and updated ed. (Freeman, San Francisco, 1989). 5, 6

L. Smolin, *The Life of the Cosmos* (Oxford University Press, New York, 1997). 5

V. J. Stenger, *The Comprehensible Cosmos: Where Do the Laws of Physics Come from?* (Prometheus, Amherst, N.Y., 2006). 2, 6

———, *Physics and Psychics: The Search for a World beyond the Senses* (Prometheus, Amherst, N.Y., 1992). 1, 6

J. S. Trefil, *Reading the Mind of God: In Search of the Principle of Universality* (Charles Scribner's Sons, New York, 1989). 2, 5, 6, 8

———, *The Moment of Creation: Big Bang Physics from Before the First Millisecond to the Present Universe* (Charles Scribner's Sons, New York, 1983). 4–6

P. van Inwagen, *Metaphysics*, 3rd ed. (Westview, Boulder, Colo., 2008). 4, 8

S. Weinberg, *Facing Up: Science and Its Cultural Adversaries* (Harvard University Press, Cambridge, Mass., 2001). 1

———, *Dreams of a Final Theory* (Pantheon Books, New York, 1993). 5

————, *The First Three Minutes: A Modern View of the Origin of the Universe* (Basic Books, New York, 1977). 5

A. Zee, *Fearful Symmetry: The Search for Beauty in Modern Physics* (Macmillan, New York, 1986, and Princeton University Press, Princeton, N.J., 2007). 3, 6, 8

INDEX

context, invariant, 128–30, 137
Copernicus, Nicolaus, 87
correlation, 2, 10, 77, 92; quantum, 101, 117, 134, 157. *See also* quantum nonseparability
cosmogony, 76
cosmological evolution, 23, 29, 49, 73, 77, 79–81, 106, 114
cosmological scheme, 29–30, 75–81, 123, 156; big-bang, 74, 79, 81; oscillatory, 73
cosmology, 23, 75–77, 81, 156
creationism, 22–23, 25, 49, 61, 123
creation science, 62

Darwin, Charles Robert, 4, 123, 130
Darwinian evolution, 4–7, 24, 27, 88–89, 123, 130, 146
data: from experiment, 31–32, 36–40, 42, 90–91; interpretation of, 76–79, 123; from observation, 12, 31, 40, 42, 55, 75–80, 90, 107, 123, 141, 145, 155
deconstruction, 6–7
deity. *See* god
Descartes, René, 2
design, intelligent, 23, 49, 123
dimension, 134–37; extra, 49–50
Dirac, Paul Adrien Maurice, 53–54
Doppler, Christian Johann, 79

economy, 66, 143–44, 146, 149, 151, 158
Einstein, Albert, 51, 79, 100, 111
electricity, 53, 63, 96, 100–101, 104
electrochemistry, 1, 63
electromagnetism, 9, 21
electron, 53–54, 97, 143
elementary particles, 34, 53, 77, 80, 96, 105, 112; properties of, 49, 77, 80
event, 61, 102; anomalous, 69–70, 115, 133; submicroscopic, 114
evidence, 3, 13, 16–19, 23–24, 31, 34, 59, 87, 117

evolution, 110, 124; biological, 23, 61, 123, 133; cosmological, 23, 29, 49, 73, 77, 79–81, 106, 114; Darwinian, 4–7, 24, 27, 88–89, 123, 130, 146; law of, 102–106, 117, 157; natural, 102–106, 117
existence, objective, 3–4, 14
experiment, data from, 31–32, 36–40, 42, 90–91
explanation, anthropic, 116, 124, 126–32, 136–37, 158; of space and time, 134–37, 158
explanatory chain, 130
extranaturality, 64–66, 68
extrasensory perception (ESP), 33, 69, 115

falsifiability, 47, 50, 54, 57, 80, 155
force, 51, 55–57, 99, 104–5, 107, 109; electric, 100–101, 104; gravitational, 51, 57, 98, 100–101, 104, 124, 126–29; magnetic, 100–101, 104; nuclear, 101, 124
framework, 49; conceptual, 60, 66; logical, 21, 46; philosophical, 27, 59–60, 66; of science, 70, 73; self-consistent, 20, 25
fundamentality, 59–60, 126–32; of aspects of nature, 48–49, 110, 115–16, 123–24, 127–32, 134, 137, 148; of human existence, 48, 116, 123, 127–31, 134, 137, 158; of theories, 47–50, 56–57, 126, 155

galaxy, 13, 71, 77–80, 99–100, 106, 114
Galileo, 87
generality, 47–48, 50–52, 54, 56, 57, 155; in regard to anthropic principle, 123, 126–27, 129, 137, 158
geocentric model, 13, 20, 87
god, 17–19, 49–50, 60, 62, 151
gravitational force, 51, 57, 98, 100–101, 104, 124, 126–29

laws of nature, 21, 36, 40, 75, 85, 90, 106, 114, 132, 155; hybrid position in regard to, 88–89, 94, 109, 116, 141, 156; reality of, 86; subjectivity of, 86, 88, 116, 156

level. *See* scale

life, 19, 27, 64–66, 93

logic, 3, 34, 46, 65, 76, 111, 124–25, 142

logical construct, 13–14, 24, 154

logical deduction, 20, 24, 154

logical equivalence, 126, 133

logical implication, 46–47, 50, 56–57, 125–29, 137, 155

logical inconsistency, 61, 65

logical self-consistency, 13, 20–21, 25, 76

logical truth, 20–21, 24, 46, 50, 57, 154, 155

Mach, Ernst, 100

Machian influence, 100–101

Mach principles, 100–101, 107–11, 117, 157; extended, 108–11, 115, 117

magnetism, 29, 100–101, 104

many-worlds interpretation, 72–73

mass, 51, 53, 55–56, 99

material being, 59, 79, 134–37

materialism, 27, 32

material universe, 28, 31–32, 42, 60, 62, 122, 155

mathematical relation, 20, 91

mathematical theorem, 20, 25, 154

mathematical truth, 13–14

mathematics, 13–14, 24, 50, 56

matter, 99–100, 107, 117, 157

Maxwell, James Clerk, 21; laws of, 21

measurement, 8–9, 31–32, 37, 39, 121, 135, 141–42

mechanical phenomenon, 21, 56

mental construct, 1, 14, 18, 86, 88, 116, 121, 156

metaphysical construct, 73, 151

metaphysical ensemble, 72–74

metaphysical position, 61, 86–88, 92, 116, 140–43, 151, 156

metaphysical principles, Mach and extended Mach principles as, 109–110

metaphysics, 59–61, 66, 86, 100, 116, 131–32, 140, 155–56; cosmology as branch of, 75–77, 81, 156; domain of, 60, 66, 110, 116, 155–56

metatime, 73

microscopic phenomenon, 95–96

mind, 2, 10–11, 17–18, 88, 149, 154; materiality of, 27, 32, 62–66; of observer, 86–87, 116, 141, 156. *See also* mental construct

mind stuff, 4, 10, 90

models: geocentric, 13, 20, 87; heliocentric, 13, 40, 87

molecule, 63–64, 93, 112, 114. *See also* scale: molecular

motion, 33–34, 36, 51, 53, 56–57, 99–100, 104, 107, 109–10; laws of, 21, 47–48, 50, 55–56, 99, 104; planetary, 40–42, 47–48, 50, 55–56, 95, 104, 155. *See also* Kepler: laws of; Newton: laws of

natural evolution, 102–106, 117

naturality, 48–49, 56, 115, 122–23, 128, 155

natural phenomenon, 19, 32, 36, 45, 47, 91–92, 117, 122–23, 128, 137

natural world, 17–19

nature: behavior of, 12, 18, 86, 92, 99, 107, 109, 157; complexity in, 91–92, 94–95; definition of, 31; fundamentality of aspects of, 48–49, 110, 115–16, 123–24, 127–32, 134, 137, 148; interaction of humans with the universe in definition of, 31–32, 39, 42, 59–60, 91, 150, 155; predictability in, 71, 121, 132; quantum aspect of, 134, 147–48, 150; reproducibility of aspect of, 30, 32–33, 39,

42, 70, 85, 94, 144, 155; role of
Homo sapiens in, 92, 94, 109,
122–23; simplicity in, 40, 57, 76,
91–94
nature, laws of, 21, 36, 40, 75, 85,
90, 106, 114, 132, 155; hybrid
position in regard to, 88–89, 94,
109, 116, 141, 156; reality of, 86;
role of Homo sapiens in, 92, 94,
109, 122–23; subjectivity of, 86,
88, 116, 156
necessary condition, 125
neutron, 53, 124
Newton, Isaac, 21; laws of, 21,
55–57, 72, 86, 95, 99–100, 104,
106, 108–109, 155
nonseparability, quantum, 96–97,
101, 116–17, 134, 157. *See also*
separability
nontranscendence of worldview,
61–62, 64–66, 150, 156. *See also*
transcendence
nuclear force, 101, 124
nucleus, atomic, 53, 112, 114, 124,
142

object: of belief, 11; of investigation,
98, 111, 131–32; localized, 149; of
thought, 2, 154
objectivity, 1–2, 13–15, 24, 42, 116,
154–55; definition of, 2; and
existence, 3–4, 14; and knowledge,
5–6, 8–10, 12–13, 24, 154; and
order, 88–91, 141, 116, 156; and
reality, 3, 87, 141, 143–51,
158–59; in science, 23–24, 29–32;
and truth, 15, 19–20, 22, 24–25,
54, 61, 66, 154; and the world, 1,
3–4, 9, 16–19, 22–25, 32, 86, 154.
See also anthropic principle:
objective difficulty of
observation, 31, 55, 59–60, 94–98,
121, 141, 144, 147; astronomical,
23, 40, 76, 78, 80, 113; data from,
12, 42, 75–80, 90, 107, 123, 145,
155

observer: mind of, 86–87, 116, 141,
156; and observed, 59, 94–97,
116, 157
order, 36–37, 39–40, 42, 56, 69, 74,
81, 103, 111, 132, 155, 158;
objective, 88–91, 141, 116, 156; in
quasi-isolated systems, 102, 106–7,
109, 111, 113, 117
orderlessness: of phenomena or
system, 70, 133, 157; of the
universe, 74–75, 81, 85, 111,
113–16, 132, 156
ordinary-size phenomenon, 95–96, 148
outer world, 2, 4, 10, 18, 154

parapsychological phenomenon, 33,
36, 69–70, 115, 133
particles, elementary, 34, 53, 77, 80,
96, 105, 112; properties of, 49,
77, 80
perception, 4–6, 10–11, 50–51, 88,
60, 121–22, 127, 146, 154;
extrasensory (ESP), 33, 69, 115; of
reality, 145–48, 150
phenomenon, 31, 56, 60, 96, 98, 141;
astronomical, 95, 106, 146; atomic,
53; biological, 48; fundamental,
130; geological, 71; incidental,
130; irreproducible, 39, 69–70;
mechanical, 21, 56; microscopic,
95–96; molecular, 97; natural,
19, 32, 36, 45, 47, 91–92, 117,
122–23, 128, 137; of objective
reality, 150; observed, 21, 96–97;
orderly, 88; ordinary-size, 95–96,
148; parapsychological, 33, 36,
69–70, 115, 133; real-world, 21;
reproducible, 40, 42, 69, 71, 76,
155; submicroscopic, 96, 146;
transient, 69, 115, 133; unique, 33,
72, 74, 76, 81, 85, 156; within the
universe, 28, 81, 85, 156; unpredictable, 39, 60. *See also* lawlessness: of phenomena or system;
orderlessness: of phenomena or
system

uncertainty principle, 97
unification, 51–52, 56–57, 81, 127,
    129, 137, 155, 158
unique phenomenon, 33, 72, 74, 76,
    81, 85, 156
universe: age of the, 123–24; aspect
    of the, 27, 60, 75, 81, 85, 129,
    133, 156; behavior of, 77–78, 85,
    110–11, 114; branch, 72–73;
    evolution of the, 29, 49, 73, 77,
    79, 81, 106, 114; geocentric, 13,
    87; interaction of humans with,
    49, 60, 70, 94, 121, 144; irrel-
    evance of theory to the, 69–70, 81,
    156; irreproducibility of the, 72,
    74–76, 81, 85, 156; island, 60, 74;
    lawlessness of, 74, 81, 85, 106,
    111, 113–17, 133–34, 156–57;
    material, 28, 31–32, 42, 60, 62,
    122, 155; matter in the, 99–100,
    107, 117, 157; orderlessness of,
    74–75, 81, 85, 111, 113–16, 132,
    156; origin of the, 76–77; other,
    49, 60, 72–74; phenomenon
    within the, 28, 81, 85, 156;
    properties of the, 49; state of the,
    75, 77, 80–81, 106, 158; unique,
    72, 75, 156; unpredictability of
    the, 69, 74–75, 81, 85, 113, 156;
    as a whole, 28, 48, 72, 74–75, 81,
    85, 106–9, 115–17, 156–157;
    working of the, 75–77, 81, 156.
    See also cosmology; evolution:
    cosmological; interaction: of

humans with the universe in
    definition of nature
unpredictability: of aspect, 36, 39,
    45, 60, 114; quantum, 112, 114,
    133; of the universe, 69, 74–75,
    81, 85, 113, 156. See also event:
    anomalous; phenomenon:
    transient; phenomenon: unique;
    predictability
unverifiable assumption, 75–76

world: inner, 3, 10–11, 15–16,
    21–22, 24, 154; knowledge of the,
    4; natural, 17–19; objective, 1,
    3–4, 9, 16–19, 22–25, 32, 86, 154;
    outer, 2, 4, 10, 18, 154; pushing
    against, 3–4; real, 2–5, 7, 9–19,
    20–21, 23, 24–25, 154; subjective,
    3, 10, 15, 21, 24. See also
    many-worlds interpretation
worldview, 48, 50–51, 56, 60, 65–66,
    86–88, 116, 156; holist, 92, 116,
    131, 156; hybrid, 88; idealist, 87;
    nontranscendent, 61–62, 64–66,
    150, 156; realist, 86, 141;
    reductionist, 156; subjectivity of,
    65, 94, 156; transcendent, 61–62,
    64, 66, 123, 151, 156, 159. See
    also holism; hybrid position in
    regard to laws of nature; idealism;
    metaphysical position; positivism;
    realism; reductionism; religion;
    solipsism